HOLT BIOSOU
LAB PROG

MW01181457

BIOTECHNOLOGY

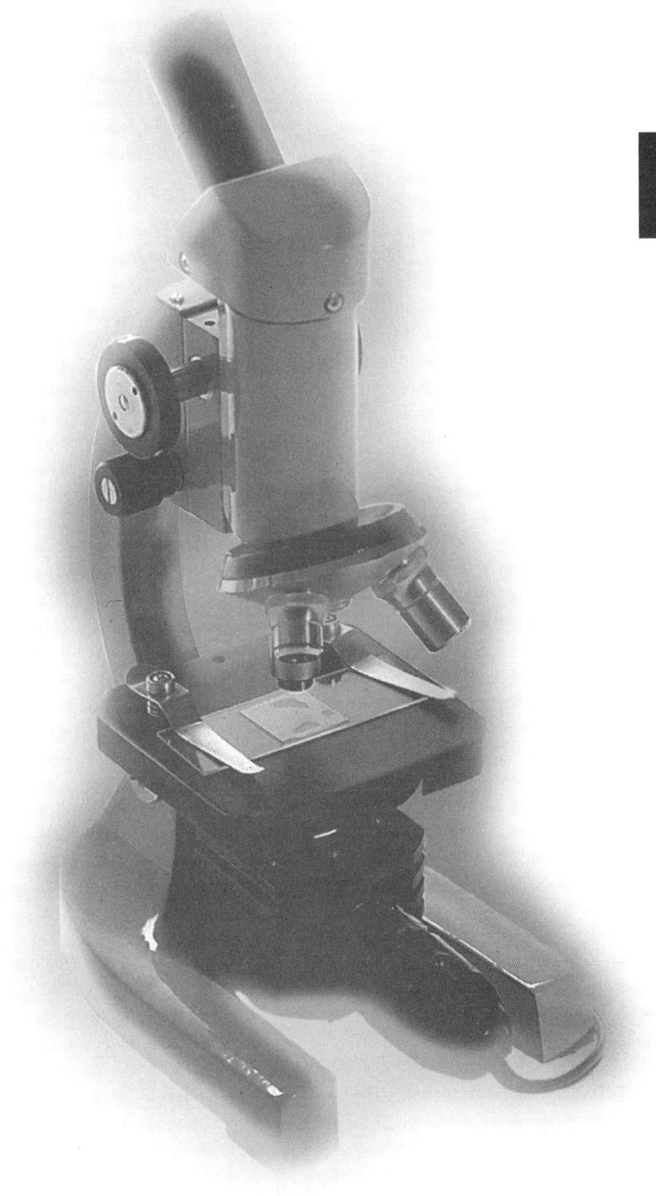

INCLUDES
LABS D1–D12

HOLT, RINEHART AND WINSTON
Harcourt Brace & Company
Austin • New York • Orlando • Atlanta • San Francisco • Boston • Dallas • Toronto • London

BIOTECHNOLOGY

Staff Credits

Editorial Development
Carolyn Biegert
Janis Gadsden
Debbie Hix

Copyediting
Amy Daniewicz
Denise Haney
Steve Oelenberger

Prepress
Rose Degollado

Manufacturing
Mike Roche

Design Development and Page Production
Morgan-Cain & Associates

Acknowledgments

Contributors
David Jaeger
Will C. Wood High School
Vacaville, CA

George Nassis
Kenneth G. Rainis
WARD'S Natural Science Establishment
Rochester, NY

Suzanne Weisker
Science Teacher and Department Chair
Will C. Wood High School
Vacaville, CA

Editorial Development
WordWise, Inc.

Cover
Design—Morgan-Cain & Associates
Photography—Sam Dudgeon

Lab Reviewers

Lab Activities
Ted Parker
Forest Grove, OR

Mark Stallings, Ph.D.
Chair, Science Department
Gilmer High School
Ellijay, GA

George Nassis
Kenneth G. Rainis
Geoffrey Smith
WARD'S Natural Science Establishment
Rochester, NY

Lab Safety
Kenneth G. Rainis
WARD'S Natural Science Establishment
Rochester, NY

Jay Young, Ph.D
Chemical Safety Consultant
Silver Spring, MD

Printed in the United States of America

ISBN 0-03-051407-X

1 2 3 4 5 6 022 00 99 98 97

Contents

Organizing Laboratory Data

Your data are all the records you have gathered from an investigation. The types of data collected depend on the activity. Data may be a series of weights or volumes, a set of color changes, or a list of scientific names. No matter which types of data are collected, all data must be treated carefully to ensure accurate results. Sometimes the data seem to be wrong, but even then, they are important and should be recorded accurately. Remember that nature cannot be wrong, regardless of what you discover in the laboratory. Data that seem to be "wrong" are probably the result of experimental error.

There are many ways to record and organize data, including data tables, charts, diagrams, and graphs. Your teacher will help you decide which format is best suited to the type of data you collect.

It is important to include the appropriate units when you record data. Remember that data are measurements or observations, not merely numbers. Data tables, graphs, and diagrams should have titles that are descriptive and complete enough to ensure that another person could understand them without having been present during the investigation.

Many important scientific discoveries have been made accidentally in the course of an often unrelated laboratory activity. Scientists who keep very careful and complete records sometimes notice unexpected trends in and relationships among data long after the work is completed. The laboratory notebooks of working scientists are studded with diagrams and notes; every step of every procedure is carefully recorded.

Data Tables and Charts

Data tables are probably the most common means of recording data. Although prepared data tables are often provided in laboratory manuals, it is important that you be able to construct your own. The best way to do this is to choose a title for your data table and then make a list of the types of data to be collected. This list will become the headings for your data columns. For example, if you collected data on plant growth that included both the length of time it took for the plant to grow and the amount of growth, you could record your data in a table like this:

Plant Growth Data

These data are the basis for all your later interpretations and analyses. You can always ask new questions about the data, but you cannot get new data without repeating the experiment.

Graphs

After data are collected, you must determine how to display them. One way of showing your result is to use a graph. Two types of graphs are commonly used: the line graph and the bar graph. In a line graph, the data are arranged so that two variables are represented as a single point. You could easily make a line graph of the data shown in the growth table. The first step is to draw and label the axes. Before you do this, however, you must decide which column of data should be represented on the *x*-axis (horizontal axis) and which should be represented on the *y*-axis (vertical axis).

Experiments have two types of variables, or factors that may change. Independent variables are variables that could be present even if other factors were not. For the example above, "Time" is an independent variable because time exists regardless of whether plants are present. An independent variable is, by convention, plotted on the *x*-axis of a graph. Dependent variables are variables that change because an independent variable changes. A dependent variable for this example would be "Height of plant." Dependent variables are plotted on the *y*-axis of a graph.

Next you must choose the scale for the axes of your graph. You want the graph to take up as much of the paper as possible because large graphs are much easier to read and make than small ones. For each axis, you must choose a scale that uses the largest amount of graph paper. Remember, once you choose the interval for the scale (the number of days each block represents on the *x*-axis, for example), you cannot change it. You cannot say that block one represents 1 day and block two represents 10 days. If you change the scale, your graph will not accurately represent your data.

Plant Growth

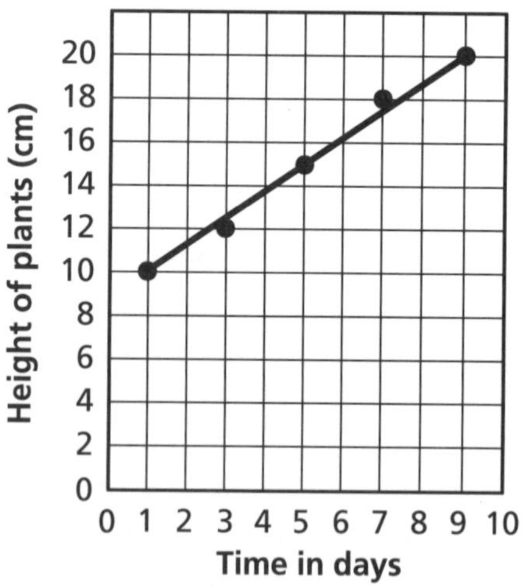

The next step is to mark the points for each pair of numbers. When all points are marked, draw the best straight or curved line between them. Remember that you do not "connect the dots" when you draw a graph. Instead, you should draw a "best fit" curve—a line or smooth curve that intersects or comes as close as possible to your set of data points.

If you choose to represent your data by using a bar graph, the first steps are similar to those for the line graph. You must first choose your axes and label them. The independent variable is plotted on the *x*-axis, and the dependent variable is plotted on the *y*-axis. However, instead of plotting points on the graph, you represent the dependent variable as a bar extending from the *x*-axis to where you would have drawn the points. Using the sample data on plant growth, for example, on day 1, the height of the plant was 10 cm. On your graph, you would make a bar that extends to the height of the 10 cm mark on the *y*-axis.

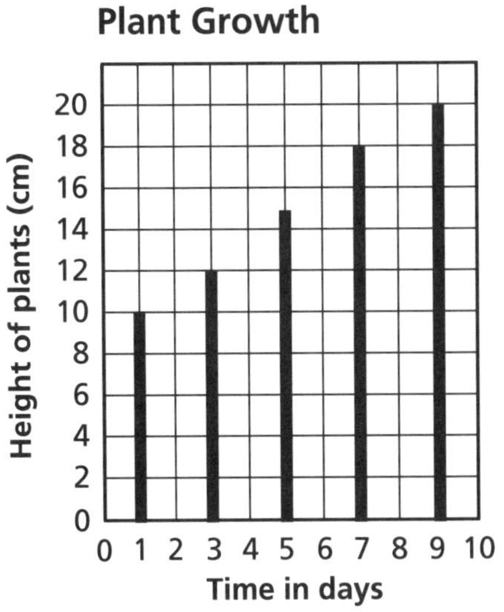

Diagrams

In some cases, the data you must represent are not numerical. That means that they cannot be put into a data table or graphed. The best way to represent this type of information is to draw and label it. To do this, you simply draw what you see and label as many parts or structures as possible. This technique is especially useful in the biology laboratory, where many investigations involve the observation of living or preserved specimens. Remember, you do not have to be an artist to make a good laboratory drawing.

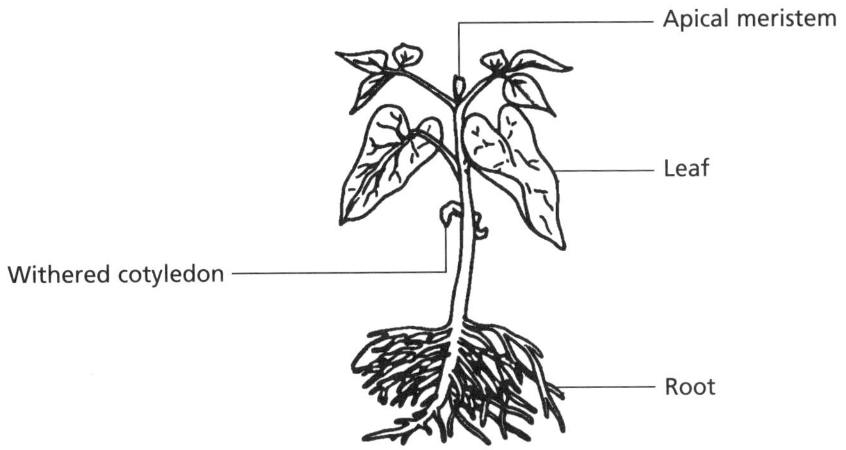

Bean seedling

There are several things you need to remember as you make your laboratory drawing. First, make the drawing large enough to be easily studied. Include all of the visible structures in your drawing. Second, drawings should also show the spacing between the parts of the specimen in proportion to its actual appearance. Size relationships are important in understanding and interpreting observations. Third, in order for your drawing to be the most useful to you, you need to label it. All labels should be clearly and neatly printed. Lines drawn from labels to the corresponding parts should be straight, so be sure to use a ruler. Label lines should never cross each other. Finally, be sure to title the drawing. Someone who looks at your drawing should be able to identify the specimens. Remember, neatness and accuracy are the most important parts of any laboratory drawing.

Laboratory Safety

Your biology laboratory is a unique place where you can learn by doing things that you couldn't do elsewhere. It also involves some dangers that can be controlled if you follow these safety notes and all instructions from your teacher.

It is your responsibility to protect yourself and other students by conducting yourself in a safe manner while in the laboratory. Familiarize yourself with the printed safety symbols—they indicate additional measures that you must take.

While in the Laboratory, at All Times . . .

- **Familiarize yourself with a lab activity—especially safety issues—before entering the lab.** Know the potential hazards of the materials, equipment, and the procedures required for the activity. Ask the teacher to explain any parts you do not understand before you start.
- **Never perform any experiment not specifically assigned by your teacher.** Never work with any unauthorized material.
- **Never work alone in the laboratory.**
- **Know the location of all safety and emergency equipment used in the laboratory.** Examples include eyewash stations, safety blankets, safety shower, fire extinguisher, first-aid kit, and chemical-spill kit.
- **Know the location of the closest telephone,** and be sure there is a posted list of emergency phone numbers, including poison control center, fire department, police, and ambulance.
- **Before beginning work: tie back long hair, roll up loose sleeves, and put on any personal protective equipment as required by your teacher.** Avoid or confine loose clothing that could knock things over, ignite from a flame, or soak up chemical solutions.
- **Report any accident, incident, or hazard—no matter how trivial—to your teacher immediately.** Any incident involving bleeding, burns, fainting, chemical exposure, or ingestion should also be reported to the school nurse or physician.
- **In case of fire, alert the teacher and leave the laboratory.**
- **Never eat, drink, or apply cosmetics.** Never store food in the laboratory. Keep your hands away from your face. Wash your hands at the conclusion of each laboratory activity and before leaving the laboratory. Remember that some hair products are highly flammable, even after application.
- **Keep your work area neat and uncluttered.** Bring only books and other materials that are needed to conduct the experiment.
- **Clean your work area at the conclusion of the lab as your teacher directs.**

- **When called for, use the specific safety procedures below.**

Eye Safety

- **Wear approved chemical safety goggles as directed.** Goggles should always be worn whenever you are working with a chemical or chemical solution, heating substances, using any mechanical device, or observing a physical process.
- **In case of eye contact**
  (1) Go to an eyewash station and flush eyes (including under the eyelids) with running water for at least 15 minutes.
  (2) Notify your teacher or other adult in charge.
- **Wearing contact lenses for cosmetic reasons is prohibited in the laboratory.** Liquids or gases can be drawn up under the contact lens and into direct contact with the eyeball. If you must wear contact lenses prescribed by a physician, inform your teacher. You must wear approved eye-cup safety goggles—similar to goggles individuals wear when swimming underwater.
- **Never look directly at the sun through any optical device or lens system, or gather direct sunlight to illuminate a microscope.** Such actions will concentrate light rays that will severely burn your retina, possibly causing blindness!

Electrical Supply

- **Never use equipment with frayed cords.**
- **Ensure that electrical cords are taped to work surfaces** so that no one will trip and fall and so that equipment can't be pulled off the table.
- **Never use electrical equipment around water or with wet hands or clothing.**

Clothing Protection

- **Wear an apron or lab coat when working in the laboratory to prevent chemicals or chemical solutions from coming in contact with skin or contaminating street clothes.** Confine all loose clothing and long jewelry.

Animal Care

- **Do not touch or approach any animal in the wild.** Be aware of poisonous or dangerous animals in any area where you will be doing outside fieldwork.
- **Always obtain your teacher's permission before bringing any animal (or pet) into the school building.**
- **Handle any animal only as your teacher directs.** Mishandling or abuse of any animal will not be tolerated!

Sharp Object Safety

- **Use extreme care with all sharp instruments, such as scalpels, sharp probes, and knives.**
- **Never use double-edged razor blades in the laboratory.**
- **Never cut objects while holding them in your hand.** Place objects on a suitable work surface.

Chemical Safety

- **Always wear appropriate personal protective equipment.** Safety goggles, gloves, and an apron or lab coat should always be worn when working with any chemical or chemical solution.
- **Never taste, touch, or smell any substance or bring it close to your eyes, unless specifically told to do so by your teacher.** If you are directed by your teacher to note the odor of a substance, do so by waving the fumes toward you with your hand. Never pipet any substance by mouth; use a suction bulb as directed by your teacher.
- **Always handle any chemical or chemical solution with care.** Check the label on the bottle and observe safe-use procedures. Never return unused chemicals or solutions to their containers. Return unused reagent bottles or containers to your teacher. Store chemicals according to your teacher's directions.
- **Never mix chemicals** unless specifically told to do so by your teacher.
- **Never pour water into a strong acid or base.** The mixture can produce heat and can splatter. Remember this rhyme:

  "Do as you oughta—
  Add acid (or base) to water."

- **Report any spill immediately to your teacher.** Handle spills only as your teacher directs.
- **Check for the presence of any source of flames, sparks, or heat (open flame, electric heating coils, etc.) before working with flammable liquids or gases.**

Plant Safety

- **Do not ingest any plant part used in the laboratory (especially seeds sold commercially).** Do not rub any sap or plant juice on your eyes, skin, or mucous membranes.
- **Wear protective gloves (disposable polyethylene gloves) when handling any wild plant.**
- **Wash hands thoroughly after handling any plant or plant part (particularly seeds).** Avoid touching your hands to your face and eyes.
- **Do not inhale or expose yourself to the smoke of any burning plant.**
- **Do not pick wildflowers or other plants unless directed to do so by your teacher.**

Proper Waste Disposal

- **Clean and decontaminate all work surfaces and personal protective equipment as directed by your teacher.**
- **Dispose of all sharps (broken glass and other contaminated sharp objects) and other contaminated materials (biological and chemical) in special containers as directed by your teacher.**

Hygienic Care

- **Keep your hands away from your face and mouth.**
- **Wash your hands thoroughly before leaving the laboratory.**
- **Remove contaminated clothing immediately; launder contaminated clothing separately.**
- **When handling bacteria or similar microorganisms, use the proper technique demonstrated by your teacher.** Examine microorganism cultures (such as petri dishes) without opening them.
- **Return all stock and experimental cultures to your teacher for proper disposal.**

Heating Safety

- **When heating chemicals or reagents in a test tube, never point the test tube toward anyone.**
- **Use hot plates, not open flames.** Be sure hot plates have an "On-Off" switch and indicator light. Never leave hot plates unattended, even for a minute. Never use alcohol lamps.
- **Know the location of laboratory fire extinguishers and fire blankets.** Have ice readily available in case of burns or scalds.
- **Use tongs or appropriate insulated holders when heating objects.** Heated objects often do not look hot. Never pick up an object with your hands unless you are certain it is cold.
- **Keep combustibles away from heat and other ignition sources.**

Hand Safety

- **Never cut objects while holding them in your hand.**
- **Wear protective gloves when working with stains, chemicals, chemical solutions, or wild (unknown) plants.**

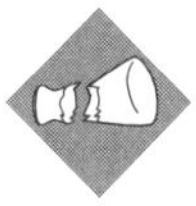

Glassware Safety

- **Inspect glassware before use; never use chipped or cracked glassware.** Use borosilicate glass for heating.
- **Do not attempt to insert glass tubing into a rubber stopper without specific instruction from your teacher.**
- **Always clean up broken glass by using tongs and a brush and dustpan.** Discard the pieces in an appropriately labeled "sharps" container.

Safety With Gases

- **Never directly inhale any gas or vapor.** Do not put your nose close to any substance having an odor.
- **Handle materials prone to emit vapors or gases in a well-ventilated area.** This work should be done in an approved chemical fume hood.

Using Laboratory Techniques and Experimental Design Labs

You will find two types of laboratory exercises in this book.

1. ***Laboratory Techniques*** labs help you gain skill in biological laboratory techniques.
2. ***Experimental Design*** labs require you to use the techniques learned in *Laboratory Techniques* labs to solve problems.

Working in the World of a Biologist

Laboratory Techniques and Experimental Design labs are designed to show you how biology fits into the world outside of the classroom. For both types of labs, you will play the role of a working biologist. You will gain experience with techniques used in biological laboratories and practice real-world work skills, such as creating a plan with available resources, working as part of a team, developing and following a budget, and writing business letters.

Tips for success in the lab

Preparation helps you work safely and efficiently. Whether you are doing a Laboratory Techniques lab or an Experimental Design lab, you can do the following to help ensure success.

- **Read a lab twice** before coming to class so you will understand what to do.
- **Read and follow the safety information** in the lab and on pages viii–xi.
- **Prepare data tables** before you come to class.
- **Record all data and observations immediately** in your data tables.
- **Use appropriate units** whenever you record data.
- **Keep your lab table organized** and free of clutter.

Laboratory Techniques Labs

Each Laboratory Techniques lab enables you to practice techniques that are used in biological research by providing a step-by-step procedure for you to follow. You will use many of these techniques later in an Experimental Design lab. The parts of a Laboratory Techniques lab are described below.

1. *Skills* identifies the techniques and skills you will learn.
2. *Objectives* tells you what you are expected to accomplish.
3. *Materials* lists the items you will need to do the lab.
4. *Purpose* is the setting for the lab.
5. *Background* is the information you will need for the lab.
6. *Procedure* provides step-by-step instructions for completing the lab and reminders of the safety procedures you should follow.
7. *Analysis* items help you analyze the lab's techniques and your data.
8. *Conclusions* items require you to form opinions based on your observations and the data you collected in the lab.

9. *Extensions* items provide opportunities to find out more about topics related to the subject of the lab.

Experimental Design Labs

Each of these labs requires you to develop your own procedure to solve a problem that has been presented to your company by a client. The procedures you develop will be based on the procedures and techniques you learned in previous Laboratory Techniques labs. You must also decide what equipment to use for a project and determine the amount you should charge the client. The parts of an Experimental Design lab are described below.

1. *Prerequisites* tells you which Laboratory Techniques labs contain procedures that apply to the Experimental Design lab.
2. *Review* tells you which concepts you need to understand to complete the lab.
3. The *Letter* contains a request from a client to solve a problem or to do a project.
4. The *Memorandum* is a note from a supervisor directing you to perform the work requested by the client, and providing clues or directions that will help you design a successful experiment or project. The *Memorandum* also contains the following: a *Proposal Checklist,* which must be completed before you start the lab; *Report Procedures,* which tells you what should be included in your lab report; *Required Precautions,* which indicate the safety procedures you should follow during the lab; and *Disposal Methods,* which tells you how to dispose of the materials used.
5. *Materials and Costs* is a list of what you might need to complete the work and the unit cost of each service and item.

What you should do before an Experimental Design lab

Before you will be allowed to begin an Experimental Design lab, you must turn in a proposal that includes the question to be answered, the procedure you will use, a detailed data table, and a list of all the proposed materials and their costs. Before you begin writing your proposal, follow these steps.

- **Read the lab thoroughly,** and jot down any clues you find that will help you successfully complete the lab.
- **Consider what you must measure or observe** to solve the problem.
- **Think about *Laboratory Techniques*** labs you have done that required similar measurements and observations.
- **Imagine working through a procedure,** keeping track of each step and of the equipment you will need.

What you should do after an Experimental Design lab

After you finish, prepare a report as described in the *Memorandum.* The report can be in the form of a one- or two-page letter to the client, plus an invoice showing the cost of each phase of the work and the total amount you charged the client. Carefully consider how to convey the information the client needs to know. In some cases, graphs and diagrams may communicate information better than words can.

Name ______________________________

Date ______________ Class ______________

D1 Laboratory Techniques: Staining DNA and RNA

Skills

- making a squash of onion root tips *(Allium sativum)*
- staining the nucleic acids of a typical plant cell
- using a compound light microscope

Objectives

- *Demonstrate* how to make a squash of onion root tips.
- *Compare* the location in the cell of DNA and RNA.

Materials

- safety goggles
- lab apron
- microscope slide
- methyl green-pyronin Y stain in dropper bottle
- forceps
- vial of pretreated *Allium* root tips
- wooden macerating stick
- watch or clock
- coverslip
- paper towels
- pencil with eraser
- compound light microscope
- prepared reference slide of plant cells
- prepared reference slide of animal cells
- mounting medium (Piccolyte II) in dropper bottle (optional)

Purpose

You have just come back from a visit to a pathology laboratory, where you observed a renal (kidney) biopsy. The pathologist wanted to determine if the person had rejected a kidney she recently received as a transplant. A methyl green-pyronin Y stain was used to see if lymphocyte-type cells were collecting around the blood vessels—a sign of possible organ rejection. These cells have RNA in their cytoplasm and when stained become bright pink. The pathologist has given you some stain to try on plant cells and mentioned that you can use the stain for detecting both DNA and RNA. You are going to see if the pathologist is correct and if it is difficult to see the difference in staining results.

Background

Root tip cells of onions (*Allium sativum*) are frequently used to study DNA and RNA in plant cells. In plants, mitosis occurs in special growth regions called **meristems** located at the tips of the roots and stems. To observe chromosomes in stem and root meristems, biologists prepare a special kind of slide called a **squash.** This preparation is just what it sounds like. Tissue containing actively dividing cells is removed from a root or stem meristem and treated with hydrochloric acid to fix the cells, or to stop them from dividing. The cells are then stained, made into a wet mount, and squashed and spread into a single layer by applying pressure to the coverslip.

The stain methyl green-pyronin Y is a mixture of two different stains. Methyl green is absorbed by DNA only and stains the DNA blue. Pyronin Y is absorbed by RNA only and stains the RNA pink. Therefore, methyl green-pyronin Y can be used to differentiate between the two nucleic acids.

Procedure

1. Put on safety goggles and a lab apron.

2. Place a microscope slide on a paper towel on a smooth, flat surface. **CAUTION: Glassware is fragile. Notify your teacher promptly of any broken glass or cuts. Do not clean up broken glass or spills unless your teacher tells you to do so.** Add two drops of methyl green-pyronin Y stain to the center of the slide. **CAUTION: Methyl green-pyronin Y stain will stain your skin and clothing. Promptly wash off spills to minimize staining.**

3. Use forceps to transfer a prepared onion root tip to the drop of stain on the microscope slide.

4. Carefully smash the root tip by gently but firmly tapping the root with the end of a wooden macerating stick. *Note: Tap the macerating stick in a straight up-and-down motion.*

5. Allow the root tip to stain for 10 to 15 minutes. *Note: Do not let the stain dry. Add more stain if necessary.*

6. Place a coverslip over your preparation, and cover the slide by folding a paper towel over it. Using the eraser end of a pencil, gently, but forcefully, press straight down (with no twisting) on the coverslip through the paper towel. Apply only enough pressure to squash the root tip into a single cell layer. *Note: Be very careful not to move the coverslip while you are pressing down with the pencil eraser. Also be very careful not to press too hard. If you press too hard, you might break the glass slide and tear apart the cells in the onion root tip.*

   ♦ Why do you squash and spread out the root tip?

   ____

   ____

   ____

   ____

7. Examine your prepared slide under both the low power and the high power of a compound light microscope. *Note: Remember that your mount is fairly thick, so be careful not to switch to the high-power objective too quickly. You may shatter the coverslip and destroy your preparation. You will need to focus up and down carefully with the fine adjustment to better see the structures under study.* Complete the data table on the next page.

Color of Cell Structures in Onion Root Tip Cells

Structure	Stained color
nucleus	
nucleolus	
cytoplasm	
chromosomes	

8. In the space below, draw and label a representative plant cell from your prepared slide. Include all visible organelles. Indicate where DNA and RNA are found in the cell.

9. Observe the prepared reference slides of plant and animal cells. Compare them with the slide you prepared.

- ♦ How does your slide compare with the prepared slide of onion root tip cells?

____

____

____

10. Dispose of your materials according to the directions from your teacher.

11. Clean up your work area and wash your hands before leaving the lab.

Analysis

12. What color did the deoxyribonucleic acid (DNA) stain in your root tip squash?

____

13. How do you know this material is DNA?

14. What color did the nucleoli appear in the stained slide? What does this tell you about the composition of nucleoli?

15. Did each nucleus have only one nucleolus or several? What appeared to be the most common number of nucleoli?

16. Were you able to see any cells in the process of mitotic division? If so, what did the cells look like?

Conclusions

17. Where is DNA located in both plant and animal cells?

18. Where is RNA located in both plant and animal cells?

Extensions

19. Make a permanent slide of your root tip squash preparation. Ask your teacher to provide you with a mounting medium (Piccolyte II). To make a permanent slide, remove the coverslip from your wet mount. Add a drop of the mounting medium, then replace the coverslip. Place the slide on a flat surface and allow it to dry for several days.

20. Use library references to research other staining techniques.

Name ______________________________

Date ____________ Class ____________

HOLT BIOSOURCES LAB PROGRAM BIOTECHNOLOGY

D2 Laboratory Techniques: Extracting and Spooling DNA

Skills

- extracting DNA from an animal cell
- spooling DNA

Objectives

- *Separate* and *collect* the DNA from bovine liver cells.
- *Describe* the appearance of DNA extracted from a cell.
- *Relate* the location of DNA in a cell to procedures for extracting it.

Materials

- safety goggles
- lab apron
- bovine liver (2 cm square)
- mortar and pestle
- fine sand
- graduated cylinder
- SDS/NaCl solution (10 mL)
- cheesecloth (several pieces, 12 cm × 12 cm)
- funnel
- test tube
- ice-water bath
- test-tube rack
- 70% ethanol (4 mL)
- inoculating loop

Purpose

You are an intern working in the city's forensics lab. You will be assisting the forensics technician with many of her routine laboratory tests and procedures. One procedure the technician does frequently is extract DNA from cells and purify it. The purified DNA is used to help prepare a DNA fingerprint to help solve crimes. To make sure you know how to do this procedure correctly, the technician has asked you to extract DNA from the cells of a piece of bovine (beef) liver and spool the DNA for observation.

Background

The extraction of DNA from cells and its purification are of primary importance to the field of biotechnology. Extraction and purification of DNA are the first steps in the analysis and manipulation of DNA that allow scientists to detect genetic disorders, produce DNA fingerprints of individuals, and even create genetically-engineered organisms used to produce beneficial products such as insulin, antibiotics, and hormones.

The process of extracting DNA, regardless of its original source, involves the following steps. The first step in extracting DNA from a cell is to **lyse,** or break open, the cell. One common way to lyse cells is to grind a piece of tissue along with a mild abrasive material in a mortar with a pestle. After the cells have been broken open, a solution containing salt (NaCl) and a detergent containing the compound SDS, or sodiumdodecyl sulfate, is used to break down and emulsify the fat and proteins that make up the cell membrane. Finally, ethanol is added. Because DNA is soluble in water, the addition of ethanol causes the DNA to **precipitate,** or settle out of solution, leaving behind all remaining cellular components that are not soluble in ethanol. Finally, the DNA can be spooled, or wound onto an inoculating loop, and pulled from the test tube.

Procedure

1. Put on safety goggles and a lab apron.
2. Place a piece of bovine liver in a mortar. Add several grains of sand.
3. Pour 10 mL SDS/NaCl solution in the mortar.
4. Use a pestle to grind the ingredients until they form a thick fluid. *Note: Be careful not to overgrind this mixture.*
5. Place several layers of cheesecloth into a funnel. Pour the contents of the mortar through the cheesecloth into a test tube until it contains at least 2 mL of the extract. *Note: You may need to gently squeeze the cheesecloth to remove all the fluid from the cheesecloth.* **CAUTION: Glassware is fragile. Notify your teacher promptly of any broken glass or cuts. Do not clean up broken glass or spills unless your teacher tells you to do so.**
6. Place the test tube in an ice-water bath.
7. Measure 4 mL of *ice-cold* ethanol in a clean graduated cylinder.
8. Hold the test tube at a 45° angle. Slowly pour the 4 mL of ice-cold ethanol into the tube. Be careful to pour the ethanol slowly down the side of the tube. *Note: Do not pour the ethanol too fast or directly into the liver solution.* As you pour the ethanol into the test tube, observe the interface line, as shown in the diagram below.

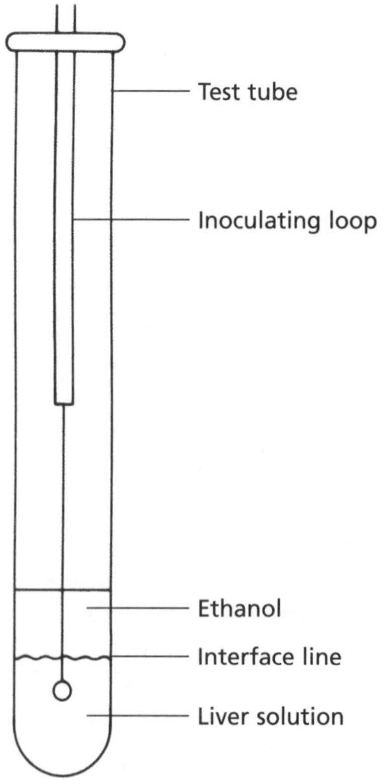

9. Gently insert an inoculating loop into the test tube as far as the interface line. Carefully and slowly move the loop in circles, as shown in the diagram at right. This motion spools the long threads of DNA around the end of the loop. *Note: Spool just enough DNA so that you can see it and observe its physical characteristics.* Lift the inoculating loop out of the solution in the test tube. While the DNA is being pulled out of the test tube, try stretching it. Then dip the inoculating loop again to get more DNA.

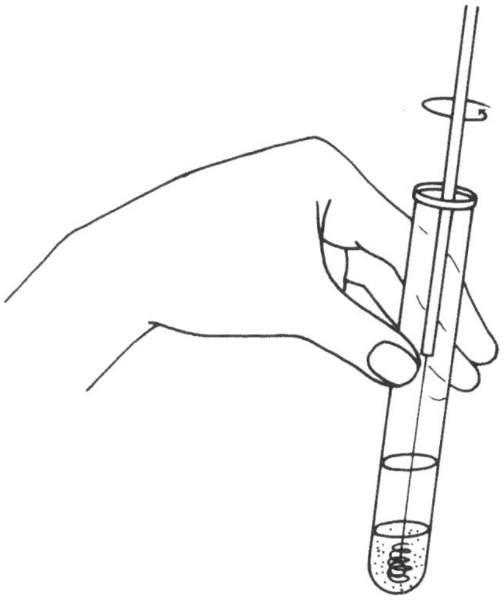

10. When spooling is complete, pull the inoculating loop from the test tube, and return the test tube to a test-tube rack.

11. Dispose of your materials according to the directions from your teacher.

12. Clean up your work area and wash your hands before leaving the lab.

Analysis

13. Describe the appearance of the DNA you spooled from the crushed bovine liver cells.

14. What was the purpose of adding sand to the liver in the mortar?

15. What happens to the cell when the SDS/NaCl solution is added to the liver mixture in the mortar?

16. What was the purpose of filtering the liver mixture through cheesecloth?

17. What was done to the DNA so that it could be observed and spooled?

18. How can you determine whether the material pulled from the test tube was DNA?

Conclusions

19. How is DNA protected inside an animal cell? How does this location relate to the procedure you used in this lab to extract DNA?

20. Biotechnologists research DNA. How could the procedure you used today facilitate their research?

Extensions

21. Find out what a DNA fingerprint is and how it is used to compare samples of DNA from different sources. Explain how the technique you learned in today's lab is used in developing a DNA fingerprint.

22. Find out about the Human Genome Project. What is this project attempting to do? How would the procedure used in this lab be used as part of the Human Genome Project?

Name ______________________________

Date ______________ Class ______________

HOLT BIOSOURCES
LAB PROGRAM
BIOTECHNOLOGY

D3 Laboratory Techniques: Genetic Transformation of Bacteria

Skills

- using aseptic technique
- predicting experimental results
- growing bacteria on agar

Objectives

- *Introduce* a plasmid into bacterial cells to genetically transform the bacteria.
- *Evaluate* whether bacterial cells were transformed by observing the characteristics of bacteria grown on petri dishes that contain agar with X-gal.

Materials

- safety goggles
- gloves
- lab apron
- disinfectant solution in squeeze bottle
- paper towels
- nontoxic permanent marker
- 15 mL sterile plastic tubes with lids (2)
- ice bath
- 1 mL sterile plastic serological pipets with 0.01 mL graduations (6)
- pipet bulb
- 0.5 mL of cold 0.5 M $CaCl_2$
- disposable inoculating loops (2)
- stock culture of *E. coli* JM101
- sterile graduated pipets (4)
- 0.01 mL of plasmid pUC8 in a 1.5 mL microtube
- petri dishes with Luria broth agar (2)
- petri dishes with Luria broth + X-gal agar (2)
- plastic foam tube holder
- 42°C water bath
- test-tube rack
- 5 mL of Luria broth
- cotton swabs (4)
- transparent tape or parafilm
- incubator
- biohazard waste container

Purpose

You are a geneticist who is involved in researching the cause of lactose intolerance, which is a digestive deficiency in humans characterized by the inability to digest lactose, a milk sugar. You decide to conduct an experiment using bacterial cells that cannot metabolize lactose. You design an experiment to genetically change the bacteria so they can digest lactose. Today you will test your experimental design to determine if it works.

Background

Escherichia coli, or ***E. coli,*** is a common bacterium found in the intestines of many mammals. *E. coli* can be classified as lac+ or lac− depending upon its ability to digest lactose. The enzyme **β-galactosidase** breaks down lactose into glucose and galactose for energy consumption. If *E. coli* produces β-galactosidase, it can metabolize lactose and is **lac+**. If *E. coli* is not capable of producing β-galactosidase, it cannot metabolize lactose and is **lac−**. Cells of the JM101 strain of *E. coli* are not capable of producing β-galactosidase and are lac−.

Bacteria normally have two types of DNA, a main chromosome and a circular DNA molecule called a **plasmid.** Plasmids contain only a few genes and are used in genetic engineering to insert genes into other organisms. Genetic engineers refer to plasmids by code numbers. The plasmid pUC8 contains genetic instructions for synthesizing β-galactosidase.

Genetic transformation is the process of changing an organism by transferring genetic material from another organism. The plasmid pUC8 contains DNA that will transform *E. coli* JM101 cells into cells that can metabolize lactose. Bacterial cells are more permeable to DNA (more likely to allow DNA to pass through their cell walls and membranes) if they are treated with a calcium chloride solution and exposed to low and high temperatures. The sudden temperature change creates a flow into the bacterial cell, bringing the plasmid into the bacterial cell. The DNA from the plasmid flows through the bacterial cell membrane and wall more easily after treatment.

The compound **X-galactoside,** or **X-gal,** is metabolized by β-galactosidase in a manner similar to the metabolism of lactose. A waste product of X-gal metabolism is a bright blue color. If β-galactosidase is present in bacterial cells, X-gal is metabolized and a bright blue ring forms around the colonies. *E. coli* has been transformed from lac− to lac+ if the colonies turn bright blue after growing on a nutrient medium treated with X-gal.

Procedure

Part 1—Transformation Technique

1. Predict the results of a successful transformation by placing a check mark (✔) in the appropriate spaces in the table below. In the table, the presence of or absence of the plasmid pUC8 is shown with a "+" or a "−," respectively.

Dish	Conditions under which *E. coli* JM101 is grown	Evidence of bacterial growth	Evidence of X-gal digestion
Dish 1	+pUC8		
Dish 2	+pUC8 and X-gal		
Dish 3	−pUC8		
Dish 4	−pUC8 and X-gal		

2.  Put on safety goggles, gloves, and a lab apron.

3.  Use aseptic technique throughout this lab. Clean the lab-table surface with disinfectant solution and paper towels.

4. Using a permanent marker, label the lids of two 15 mL plastic tubes with the initials of everyone in your lab group. Then write +pUC8 on one lid and −pUC8 on the other lid. Place the unopened tubes in an ice bath. The tubes, bacteria, and plasmid must always be kept on ice unless the instructions state otherwise.

5. Using a 1 mL sterile plastic serological pipet, transfer 0.25 mL of cold 0.5 M $CaCl_2$ into each 15 mL plastic tube. **CAUTION: If you get a chemical on your clothing, wash it off at the sink while calling to your teacher.** Place the lids on the tubes securely, and immediately place them in the ice bath.

6. Take one plastic tube from the ice bath, and use a disposable inoculating loop to transfer several colonies from the stock culture of *E. coli* into the tube. Vigorously tap the inoculating loop against the wall of the tube to dislodge the cell mass. Then use a sterile plastic graduated pipet to gently mix the bacteria with the $CaCl_2$ solution, making sure the bacteria are suspended and that no cell mass is left on the side of the tube. *Note: Be careful not to transfer any agar. Impurities in agar can inhibit transformation.* Using a new inoculating loop and sterile pipet, repeat this process with the second tube. Check to make sure that the suspension is homogenous and that the lid on each tube is secure. Return each tube to the ice bath when the transfer is complete, and incubate the tubes on ice for 30 minutes.

7. Using a 1 mL sterile plastic serological pipet, add 0.01 mL of pUC8 to the +pUC8 tube *only. Very gently* tap the tube with your finger to mix the plasmid into the cell suspension, and return the tube to the ice bath. Leave both tubes in the ice bath for 20 minutes.

8. While the tubes are in the ice bath, obtain two petri dishes labeled *LB* (with Luria broth agar) and two petri dishes labeled *LB + X-gal* (with Luria broth + X-gal agar). Label the bottoms of the four petri dishes, as seen in the drawing at right, with the initials of each group member and the information below:

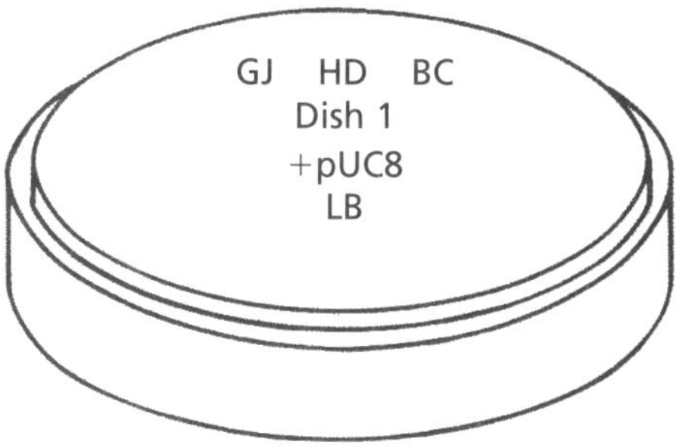

Dishes labeled *LB:*	Dishes labeled *LB + X-gal:*
Dish 1 +pUC8	Dish 2 +pUC8
Dish 3 –pUC8	Dish 4 –pUC8

9. After 20 minutes, remove the tubes from the ice bath and place each tube in a hole in a piece of plastic foam. Heat shock the tubes by floating them in a 42°C water bath for 60 seconds.

10. Remove the tubes after 90 seconds, and immediately place them back in the ice bath for 2 minutes. After 2 minutes, remove the tubes from the ice bath, and place them in a test-tube rack to return to room temperature.

11. Using a sterile plastic graduated pipet, aseptically transfer 2.5 mL of Luria broth into one tube. Gently tap the tube to mix the contents. Repeat for the second tube. Incubate the tubes at 37°C for 30 minutes.

12. Using a 1 mL sterile plastic serological pipet, place 0.25 mL from the tube labeled +pUC8 onto Dish 1. Use a new serological pipet to place 0.25 mL from the tube labeled +pUC8 onto Dish 2. Immediately spread the solution evenly over the agar in Dish 1 using a sterile cotton swab, as seen in the diagram at right. Repeat the procedure for Dish 2. Dispose of the cotton swabs and pipets according to the directions from your teacher.

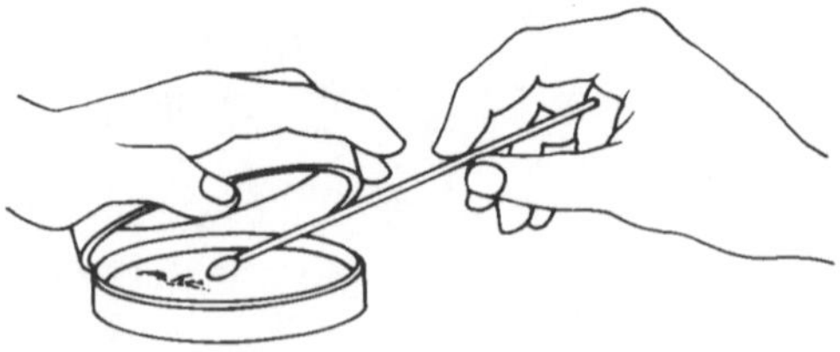

13. Repeat step 12 using the −pUC8 tube, Dish 3, and Dish 4.

14. Allow the petri dishes to set for 10 minutes or until the liquid has been completely absorbed by the agar.

15. Seal each petri dish using two pieces of transparent tape or parafilm. Then invert the dishes, stack them, and tape them together. Place the dishes in an incubator at 37°C for 24 to 48 hours.

Part 2—Observations

16. Clean the lab-table surface with disinfectant solution and paper towels.

17. After 24 hours, untape the stack of four dishes. *Note: Do not remove the tape from the individual petri dishes.* Observe each dish, and record the number and description of the colonies in the data table below. The growth of off-white colonies indicates the presence of bacteria. The presence of colonies marked with blue indicates that X-gal has been digested. *Note: Do not open sealed petri dishes.*

Results of Genetic Transformation With pUC8

Dish	Conditions	Description of results
Dish 1	+pUC8, LB	
Dish 2	+pUC8, LB + X-gal	
Dish 3	−pUC8, LB	
Dish 4	−pUC8, LB + X-gal	

18. Dispose of your materials according to the directions from your teacher.

19. Clean up your work area and wash your hands before leaving the lab.

Analysis

20. What is the significance of the blue circles?

21. Was colony growth the same for each petri dish? What can you conclude from this?

22. What part of the lab showed that the original *E. coli* bacteria were not able to digest the X-gal?

Conclusions

23. In this exploration, you created a new form of *E. coli* by placing a plasmid into a bacterial cell. Did the experimental design work? What applications does transformation have to society?

24. Today, many diabetics take human insulin that is made by bacteria. How do you think bacteria can be made to do this?

Extensions

25. *Biotechnologists* manage biological systems for human benefit. The career of a biotechnologist has important applications in medicine, food technology, and agriculture. Explore the field of biotechnology, and find out about the training and skills required to become a biotechnologist.

26. Do research to find out what laws regulate the manufacturing of new organisms and the impact biotechnology has had on science.

Name ______________________________

Date ________________ Class ________________

HOLT BIOSOURCES
LAB PROGRAM
BIOTECHNOLOGY

D4 Experimental Design: Genetic Transformation—Antibiotic Resistance

Prerequisites

- **Biotechnology D3—Laboratory Techniques: Genetic Transformation of Bacteria** on pages 9–14

Review

- cellular respiration
- procedures to induce genetic information
- screening for a characteristic

NIH
National Institutes of Health
Washington, D.C.

March 5, 1998

Caitlin Noonan
Research and Development Division
BioLogical Resources, Inc.
101 Jonas Salk Dr.
Oakwood, MO 65432-1101

Dear Ms. Noonan,

I am a researcher at the National Institutes of Health, or as it is frequently called, the NIH. A recent outbreak of bacterial infections in several European countries has caused us some concern. Because these countries are frequented by American tourists, it is likely that the bacteria causing these infections will soon reach the United States, if they haven't already. The problem with these bacteria is their apparent resistance to certain antibiotics. An outbreak of resistant bacteria could be very difficult to control.

We have been successful in isolating a plasmid from one of these strains of bacteria. The plasmid appears to have a gene for tetracycline resistance. We are asking several research companies to perform experiments with the plasmid so that we can be sure of what we are dealing with. If our suspicions are confirmed, we will take further steps to prepare for a potential outbreak.

I will be contacting you soon to discuss your participation and to provide you with more information. We look forward to your results.

Sincerely,

Lane Conn, M.D.

Lane Conn, M.D.
Director of Research
NIH

BioLogical Resources, Inc. Oakwood, MO 65432-1101

M E M O R A N D U M

To: Team Leader, Genetic Engineering Dept.

From: Caitlin Noonan, Director of Research and Development

Please review the attached letter. Dr. Conn has sent a sample of the plasmid to which she refers in her letter. Have your research team use this plasmid to genetically transform the RRI strain of *E. coli,* and screen the resulting bacterial colonies for tetracycline resistance. Give this project your top priority. It is important that we finish this project as quickly as possible, but I do not want you to sacrifice quality for speed. I am sure that you understand the importance of a contract with the National Institutes of Health. Take extreme care in providing accurate results.

I recently received word that OSHA will be inspecting us for safety standards over the next few days. I do not expect this to affect your performance; I am certain that you already follow all safety and disposal guidelines as strictly as possible.

Proposal Checklist

Before you start your work, you must submit a proposal for my approval. **Your proposal must include the following:**

_____ • the **question** you seek to answer

_____ • the **procedure** you will use

_____ • a detailed **data table** for recording observations

_____ • a complete, itemized list of proposed **materials** and **costs** (including use of facilities, labor, and amounts needed)

Proposal Approval: ____

(Supervisor's signature)

Report Procedures

When you finish your analysis, prepare a report in the form of a business letter to Dr. Conn. **Your report must include the following:**

_____ • a paragraph describing the **procedure** you followed to complete a genetic transformation of the bacteria

_____ • a complete **data table** pooling data from all research groups

_____ • your **conclusions** about whether the plasmid is tetracycline-resistant

_____ • a detailed **invoice** showing all materials, labor, and the total amount due

Safety Precautions

- Wear safety goggles, disposable gloves, and a lab apron.
- Wear oven mitts when handling hot objects.
- Glassware is fragile. Notify your teacher promptly of any broken glass or cuts. Do not clean up broken glass or spills unless your teacher tells you to do so.
- Never use electrical equipment around water, or with wet hands or clothing. Never use equipment with frayed cords.
- Wash your hands before leaving the laboratory.

Disposal Methods

- Dispose of waste material according to instructions from your teacher.
- Place solid, uncontaminated materials in a trash can.
- Place broken glass, unused plasmid, unused nutrient broth, unused bacteria, pipets, cotton swabs, inoculating loops, petri dishes, and other contaminated materials in the separate containers provided.
- Wash reusable materials such as glassware and lab utensils, and return them to the supply area.

FILE: NIH

MATERIALS AND COSTS (Select only what you will need. No refunds.)

I. Facilities and Equipment Use

Item	Rate	Number	Total
facilities	$480.00/day		
personal protective equipment	$10.00/day		
incubator	$20.00/day		
clock or watch with second hand	$10.00/day		
beaker	$5.00/day		
hot plate	$15.00/day		
thermometer	$5.00/day		
test-tube rack	$5.00/day		

II. Labor and Consumables

Item	Rate	Number	Total
labor	$40.00/hour		
test plasmid	provided		
stock culture of *E. coli* RRI	$30.00 each		
nutrient broth	$1.00/mL		
petri dish w/ +T agar	$5.00 each		
petri dish w/ −T agar	$5.00 each		
calcium chloride solution	$4.00/mL		
15 mL plastic tube w/ lid	$2.00 each		
1 mL sterile plastic pipet	$2.00 each		
sterile graduated pipet	$5.00 each		
pipet bulb	$2.00 each		
plastic-foam tube holder	$5.00 each		
disposable inoculating loops	$0.50 each		
cotton swabs	$0.10 each		
disinfectant solution	$2.00/bottle		
paper towels	$0.10/sheet		
nontoxic permanent marker	$2.00 each		
transparent tape	$0.10/m		

Fines

Item	Rate	Number	Total
OSHA safety violation	$2,000.00/incident		
		Subtotal	
		Profit Margin	
		Total Amount Due	

Name ______________________

Date ____________ Class ____________

HOLT BIOSOURCES LAB PROGRAM BIOTECHNOLOGY

D5 Laboratory Techniques: Introduction to Agarose Gel Electrophoresis

Skill

- conducting agarose gel electrophoresis

Objectives

- *Understand* the principles and practices of agarose gel electrophoresis.
- *Determine* the R_f value for each of five dye samples.
- *Demonstrate* that gel electrophoresis can be used to separate a mixture of molecules based on the charge and size of the molecules.

Materials

- safety goggles
- lab apron
- agarose gel (2.0%) on gel-casting tray
- electrophoresis system, battery-powered
- microtube rack
- microtube of bromophenol blue
- microtube of crystal violet
- microtube of orange G
- microtube of methyl green
- microtube of xylene cyanol
- microtube of dye mixture
- 10 µL micropipetter
- micropipetter tips (6)
- 250 mL graduated cylinder
- TBE running buffer (1×) (200 mL)
- 250 mL beaker
- 9 V batteries connected in series (5)
- 15 cm metric ruler

Purpose

You are an intern working in a genetic engineering laboratory. Throughout your internship, you will be assisting genetic engineers in many different experiments with DNA. In some types of DNA analysis, a sample of DNA is broken into fragments, which are then separated according to size. DNA fragments are separated by gel electrophoresis. Before you work with DNA samples, you must first learn this technique. As a training exercise, you will electrophorese five dye samples and a mixture of dyes. Like DNA fragments, the individual dyes in a mixture of dyes separate based on their size and electrical charge.

Background

Gel electrophoresis is a process that is used to separate mixtures of electrically charged molecules, such as DNA and proteins, on the basis of their size and electrical charge. The process involves passing an electric current through a **gel,** which is a slab made of a jellylike substance. Gels can be made from different substances. The gels that are commonly used to electrophorese (separate through gel electrophoresis) DNA molecules contain **agarose,** which is a sugar that comes from certain types of marine algae. Gel electrophoresis that uses gels containing agarose is called **agarose gel electrophoresis.**

During gel electrophoresis, each sample to be tested is placed in a depression called a **well,** located at one end of the gel. An electric current applied across the gel causes one end of the gel to become negative and the other end to become positive. The electric current causes the samples to migrate through small holes, or pores, in the gel. Molecules that have a negative charge migrate toward the positive electrode. Molecules with a positive charge migrate toward the negative electrode. Small molecules move more easily through the pores in a gel than do large molecules. Therefore, smaller molecules move farther and at a faster rate than larger molecules. After gel electrophoresis, the largest molecules are found closest to their wells, while the smallest molecules are found the farthest away.

Procedure

1. Put on safety goggles and a lab apron.

2. Set a micropipetter to 10 μL, and place a new tip on the end. Open the microtube containing bromophenol blue, and remove 10 μL of the dye. Carefully place the solution in the well in Lane 1 of an agarose gel. To do this, place both elbows on the lab table, lean over the gel, and slowly lower the micropipetter tip into the opening of the well before depressing the plunger. *Note: Do not jab the micropipetter tip through the bottom of the well.*

3. Using a new micropipetter tip for each tube, repeat step 2 for each of the remaining microtubes. Place each dye in a well, according to the following lane assignments:

Lane 2	Crystal violet
Lane 3	Orange G
Lane 4	Methyl green
Lane 5	Xylene cyanol
Lane 6	Dye mixture

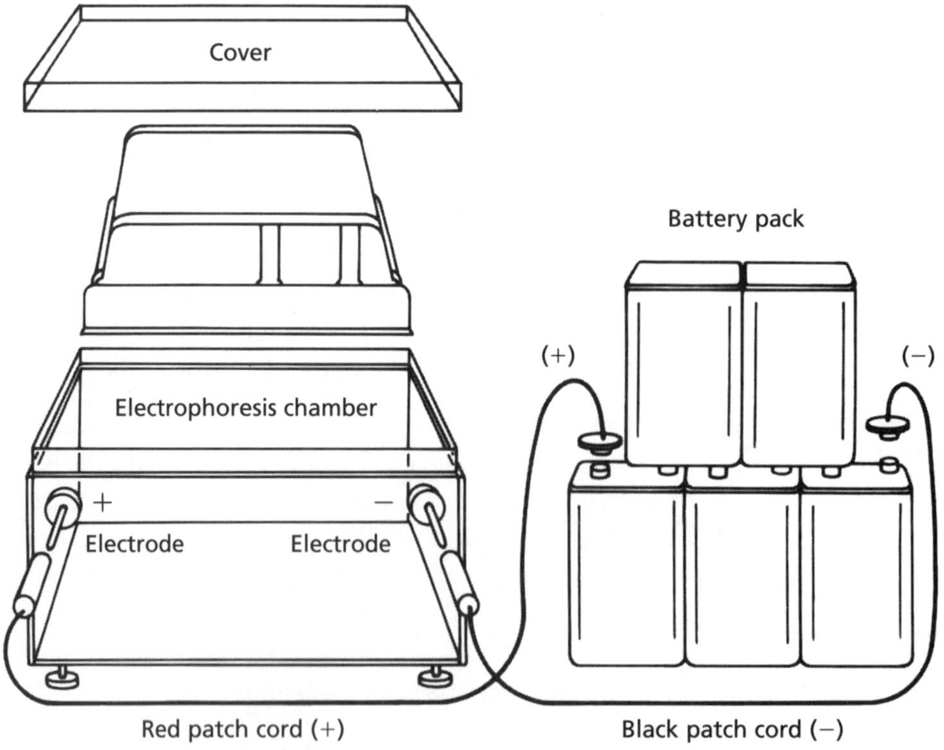

4. Carefully place the agarose gel (still in a gel-casting tray) in the electrophoresis chamber of an electrophoresis apparatus, such as the one shown above.

5. Slowly pour approximately 200 mL of 1× TBE running buffer into a beaker. **CAUTION: Glassware is fragile. Notify your teacher promptly of any broken glass or cuts. Do not clean up broken glass**

or spills unless your teacher tells you to do so. CAUTION: If you get a chemical on your skin or clothing, wash it off at the sink while calling to your teacher. If you get a chemical in your eyes, promptly flush it out at the eyewash station while calling to your teacher. Notify your teacher in the event of any chemical spill. Gently and slowly pour the running buffer from the beaker into one side of the electrophoresis chamber until the gel is completely covered (approximately 1 to 2 mm *above* the top surface of the gel). *Note: Be careful not to overfill the chamber with buffer.*

6. Place the cover on the electrophoresis chamber. Wipe off any spills around the electrophoretic apparatus before doing the next step.

7. Connect five 9 V alkaline batteries as shown in the diagram on the previous page. **CAUTION: Do not touch both ends of the patch cords or both terminals on the battery pack at the same time.** Connect the red (positive) patch cord to the red terminal on the chamber and the red terminal on the battery pack. Follow the same procedure with the black (negative) patch cord and the black terminals.

8. Observe the migration of the samples along the gel toward the red (positive) electrode and toward the black (negative) electrode.

9. Disconnect the battery pack when the dye bands in Lane 6 are fully separated and when one of the bands is near the end of the gel.

10. Remove the cover from the electrophoresis chamber. Lift the gel tray (containing the gel) from the chamber onto a piece of paper towel. Notch one side of the gel so that you can identify the lanes. *Note: The dyes are not fixed on the gel and over time (several hours) may diffuse into the gel, making the bands less distinct. Complete the next step on the same day.*

11. Use a metric ruler to measure the distance of the dye bands in Lane 6 (in mm) from each of the six sample wells. *Note: Be sure to measure from the center of the well to the center of the band.* Draw an illustration of the dye bands in each gel lane. Make your drawing in the gel diagram provided below.

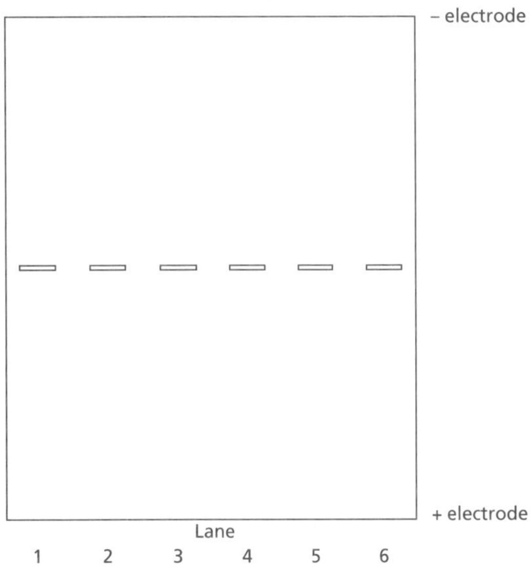

Label the dye bands in Lane 6 of your diagram A–E in order from top to bottom. Dye A will be closest to the negative pole, and Dye E will be closest to the positive pole.

12. Record the migration of each sample toward (+) or away from (−) the positive pole in the data table below. Then record the molecular charge (+ or −) of the dye in each band. Also record migration distance (in mm) for each band imaged on the gel.

Experimental Data

	Migration direction [(+) or (−)]	Molecular charge (+ or −)	Migration distance (mm)	Molecule size (BP)*
Lane 1				
Lane 2				
Lane 3				
Lane 4				
Lane 5				
Lane 6				
Dye A				
Dye B				
Dye C				
Dye D				
Dye E				

*BP stands for base-pair equivalent.

Remember that agarose gel electrophoresis is commonly used to separate mixtures of DNA molecules based on their length. The length of a DNA molecule is determined by the number of nucleotides that make up each strand of the double-stranded molecule. The two strands are held together by bonds that form between complementary nitrogen bases along each strand. These complementary pairs of nitrogen bases are called **base pairs.**

13. You can determine the relative size of each dye's molecules by using the standard curve shown in the graph on the next page. This graph, which was developed by finding the log and antilog of the distance each dye moved, gives the size of the molecules in base-pair equivalents (BP). Find the size of each dye on your gel by determining where the distance the dye moved intersects the standard curve. The number of base-pair equivalents can be read on the *y*-axis.

Standard Curve of Molecular Sizes of Dyes in BPs

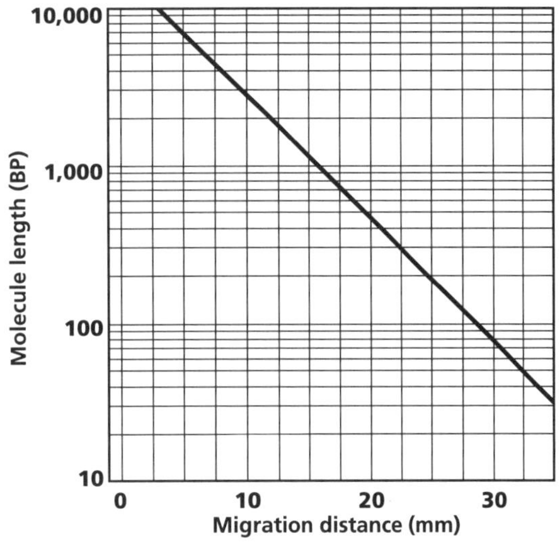

14. Dispose of your materials according to the directions from your teacher.

15. Clean up your work area and wash your hands before leaving the lab.

Analysis

16. Which dye migrated the farthest distance? the shortest distance?

____

____

17. Which dye has the largest molecules? Which dye has the smallest molecules? Explain your answer.

____

____

____

18. Why do certain dyes migrate toward the cathode and others toward the anode?

____

____

____

____

____

____

19. State some ways electrophoresis can separate molecules.

Conclusions

20. What dyes make up the mixture of dyes that you loaded into Lane 6? Identify each dye in the mixture.

21. What do you think would happen if you continued to allow electricity to run through the gel?

22. What do you think would happen if you increased the voltage (strength) of the electric current in the gel? decreased the voltage?

23. DNA and many other substances are colorless and cannot be seen as they move through the gel. How might the dyes that you electrophoresed in this lab be useful during the electrophoresis of DNA?

Extensions

24. With the help of your teacher, design an experiment to determine what concentration of agarose to use to separate molecules that are very large.

25. Find out how gel electrophoresis is applied in biological laboratories, such as those that investigate the causes of disease and that create recombinant DNA.

Name ______________________

Date ____________ Class ____________

HOLT BIOSOURCES
LAB PROGRAM
BIOTECHNOLOGY

D6 Laboratory Techniques: DNA Fragment Analysis

Skills

- conducting agarose gel electrophoresis
- calculating R_f

Objectives

- *Use* agarose gel electrophoresis to separate DNA fragments of different sizes.
- *Analyze* DNA fragments to determine their length.

Materials

- safety goggles
- lab apron
- agarose gel (0.8%) on gel-casting tray
- electrophoresis system, battery-powered
- microtube rack
- samples in labeled microtubes:
  Tube 1 "lambda DNA/*Eco*RI digest"
  Tube 2 "lambda DNA/*Hin*dIII digest"
  Tube 3 "lambda DNA/*Eco*RI/*Hin*dIII double digest"
- 10 µL micropipetter
- micropipetter tips (3)
- 250 mL graduated cylinder
- 1× TBE running buffer (200 mL)
- 250 mL beaker
- 9 V batteries (5)
- "gel handler" spatula
- staining tray
- DNA stain (100 mL per gel)
- metric ruler
- distilled water

Purpose

You are a graduate student in molecular biology. You have just gotten a summer job working on the Human Genome Project. The goal of this project is to make detailed maps showing the locations of the genes on each of the 24 different types of human chromosomes. To be able to work in the lab, you must become proficient in the following techniques: separating DNA fragments with agarose gel electrophoresis and determining the length of DNA fragments. Your first assignment is to practice these techniques with DNA from the lambda virus.

Background

Lambda is a **bacteriophage,** which is a kind of virus that can infect only bacterial cells. Lambda viruses are used in recombinant DNA technology as **vectors,** agents that carry DNA from one organism to another.

Cutting DNA into fragments is the first step in certain types of DNA analysis. The DNA is cut into fragments with a restriction enzyme. A **restriction enzyme,** or **RE,** is an enzyme that recognizes and binds to specific short sequences of base pairs in a DNA molecule and then cleaves, or cuts, the DNA at a specific site within that sequence.

Each type of restriction enzyme cuts DNA at a different base-pair sequence. The recognition sequences at which a restriction enzyme cuts a DNA molecule are relatively short, usually only four to eight base pairs in length. An RE scans the length of a DNA molecule and stops to cut the molecule only at its particular recognition site. The restriction enzyme *Eco*RI, for example, cuts DNA whenever it encounters the base-pair sequence CTTAAG.

Some restriction enzymes cut cleanly through the DNA molecule by cutting both of the complementary strands of the DNA molecule at the same position within the recognition sequence. These enzymes produce a blunt-end cut, as shown in the diagram below. The terms *5'* and *3'* refer to specific carbon atoms in the structure of the sugar in the sugar-phosphate backbone of each strand. Notice that the orientation of the 5' and 3' carbons in the two strands is opposite.

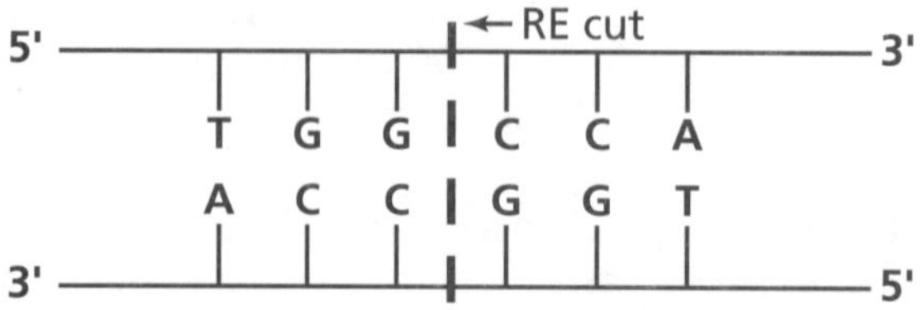

Other restriction enzymes, such as *Eco*RI and *Hin*dIII, cut the complementary strands at a different point within the recognition sequence, resulting in a staggered cut. DNA fragments that result from staggered cuts have single-stranded sections of DNA at their ends. These single-stranded ends are called **sticky ends** because two ends with complementary base sequences will join (stick together) when the complementary bases pair. The diagram below shows how *Hin*dIII cuts a DNA fragment to produce sticky ends.

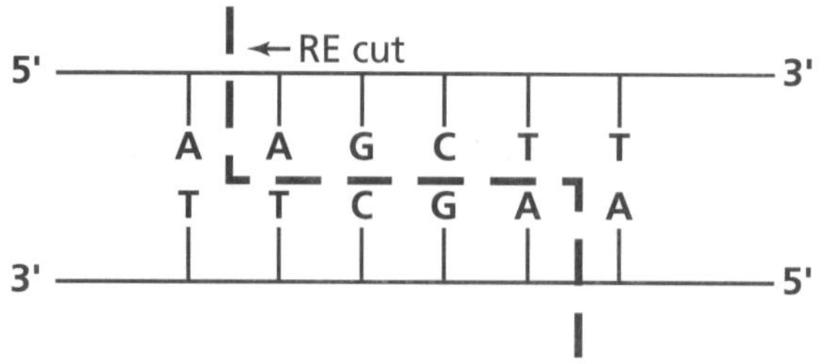

Some of the restriction enzymes that cleave DNA and the positions (base numbers in sequence) at which they cleave lambda DNA are listed in the table below. The positions in the table refer to the 5' base of the recognition sequence. Site position(s) read from left to right (5'→3' position) along the lambda DNA.

Table 1 Restriction Enzymes and Lambda DNA Cleavage Sites

Enzyme	No. of sites	Base position of cleavage sites						
		1	2	3	4	5	6	7
*Bam*HI	5	5505	22346	27972	34499	41732		
*Eco*RI	5	21226	26104	31747	39168	44972		
*Hin*dIII	7	23130	25157	27479	36895	37459	37584	44141

The fragments of DNA molecules that result when a sample of DNA is cut with restriction enzymes can be separated by agarose gel electrophoresis. During gel electrophoresis, the DNA cut by one or more restriction enzymes is loaded into a well of an agarose gel. The wells are placed near the negative electrode in a gel electrophoresis chamber. When an electric current passes through the gel, the ends of the gel become electrically charged. The DNA molecules, which are negatively charged, migrate toward the positive end of the gel.

Procedure

1. Put on safety goggles and a lab apron.

Part 1—Separating DNA Fragments

2. Set a micropipetter to 10 μL, and place a new tip on the end. Open the microtube containing lambda DNA/*Eco*RI digest (Tube 1) and remove 10 μL of the solution. Carefully place the solution into the well in Lane 1 of an agarose gel in a gel-casting tray. To do this, place both elbows on the lab table, lean over the gel, and slowly lower the micropipetter tip into the opening of the well before depressing the plunger. *Note: Do not jab the micropipetter tip into the gel.*

3. Using a new micropipetter tip for each tube, repeat step 3 for each of the remaining microtubes:
   Lane 2 (Tube 2—lambda DNA/*Hin*dIII digest)
   Lane 3 (Tube 3—lambda DNA/*Eco*RI/*Hin*dIII double digest)

4. Carefully place your loaded agarose gel (still in the gel-casting tray) in the electrophoresis chamber of an electrophoresis apparatus, such as the one shown below. Orient the gel so that the wells are closest to the black connector, or negative electrode.

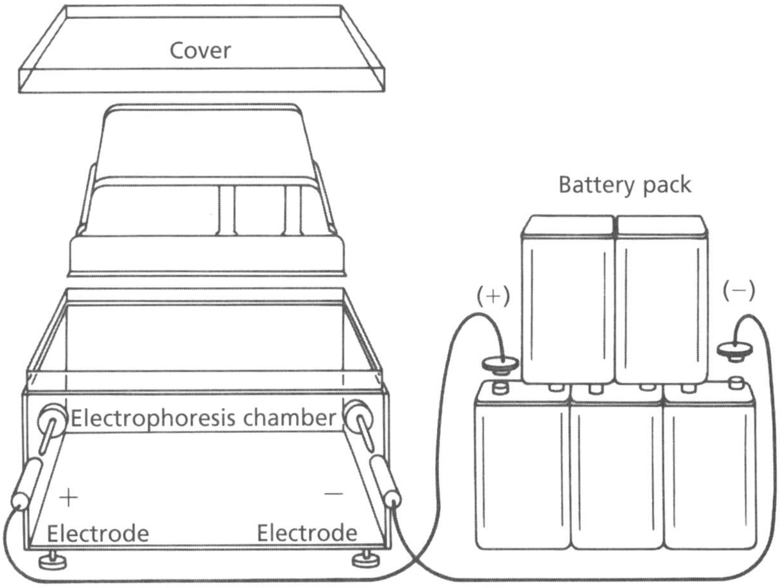

5. Slowly pour approximately 200 mL of 1× TBE running buffer into a beaker. **CAUTION: Glassware is fragile. Notify your teacher promptly of any broken glass or cuts. Do not clean up broken glass or spills unless your teacher tells you to do so. If you get a chemical on your skin or clothing, wash it off at the sink while calling to your teacher. If you get a chemical in your eyes, promptly flush it out at the eyewash station while calling to your teacher. Notify your teacher in the event of any chemical spill.** Gently and slowly pour the running buffer from the beaker into one side of the electrophoresis chamber until the gel is completely covered (1 to 2 mm *above* the top surface of the gel). *Note: Be careful not to overfill the chamber with buffer.*

6. Place the cover on the electrophoresis chamber. Wipe off any spills around the electrophoretic apparatus before doing the next step.

7. Connect five 9 V alkaline batteries as shown in the figure on the preceding page. **CAUTION: Do not touch both ends of the patch cords or both terminals on the battery pack at the same time.** Connect the red (positive) patch cord to the red terminal on the chamber and the red terminal on the battery pack. Follow the same procedure with the black (negative) patch cord and the black terminals.

8. The samples you placed in the wells contain a tracking dye. Observe the migration of the tracking dye along the gel toward the red (positive) electrode.

9. Disconnect the battery pack when the tracking dye has run near the end of the gel.

10. Remove the cover from the electrophoresis chamber. Lift the gel tray (containing the gel) from the cell, and place the gel in the staining tray. Do this by *gently* pushing the gel off the tray using a "gel handler" spatula. Notch one side of the gel so that you can identify the lanes.

Part 2—Staining a Gel

11. To stain the gel, pour approximately 100 mL of DNA stain into the staining tray until the gel is completely covered. Do not pour stain directly onto the gel. Cover and label the staining tray. Staining will be complete in 2–3 hours. **CAUTION: DNA stain will stain your skin and clothing. Promptly wash off spills to minimize staining.**

12. When the gel is stained, carefully pour the remaining stain directly into a sink. Flush down the drain with water. *Note: Do not allow the gel to move against the corner of the staining tray. The gel must remain flat; if it breaks, it will be useless.*

13. To destain the gel, add enough distilled water to the staining tray to cover the gel. *Note: Do not pour water directly onto the gel. Pour water to one side of the gel.* Let the gel sit overnight (or at least 18–24 hours). After destaining the gel, the bands of DNA will appear as purple lines against a light, almost clear background.

14. Dispose of your materials according to the directions from your teacher.

15. Clean up your work area and wash your hands before leaving the lab.

Part 3—DNA Fragment Analysis

The stained gel you produced in the steps above can be analyzed to determine the lengths of the DNA fragments. The length of DNA fragments is frequently given in nucleotide base pairs (bp) for smaller fragments and kilobase pairs (kbp) for larger ones. Smaller DNA fragments migrate through a gel faster than larger ones. The largest fragments (those with the most base pairs) remain closest to the well after electrophoresis is complete. The smallest fragments (those with the fewest base pairs) move farthest away from the well. Therefore, DNA fragments spread out in a gel lane in order from the highest to the lowest number of base pairs.

The distance a fragment moves through a gel is used to determine the fragment's length. Each fragment has a "relative mobility," or $\boldsymbol{R_f}$. The R_f value is calculated by using the following formula:

$$R_f = \frac{\text{distance the DNA fragment has migrated from the origin (gel well)}}{\text{distance from the origin to the reference point (tracking dye)}}$$

Under a given set of electrophoretic conditions (i.e., pH, voltage, time, gel type and concentration, etc.), the R_f of a DNA fragment is standard. Thus, samples containing DNA fragments of unknown length are usually electrophoresed with a marker sample, which contains fragments of known lengths. The length of an unknown DNA fragment can then be determined by comparing its R_f with those of the fragments in the DNA marker sample. One way you can do this is to use a **standard curve,** which is constructed by plotting migration distance in millimeters versus the log of fragment length.

16. Use a metric ruler to measure the distance in millimeters from the well in each lane to the tracking dye and DNA bands in the same lane. *Note: Measure from the middle of a well to the middle of a band.* As you measure the distance each DNA band moved, record this measurement under "Migration distance" in the appropriate data table on the following pages. Draw the position of the DNA bands in each lane of your gel on the "gel" diagram below. *Note: Also draw the position of the wells, based on their position on your gel.*

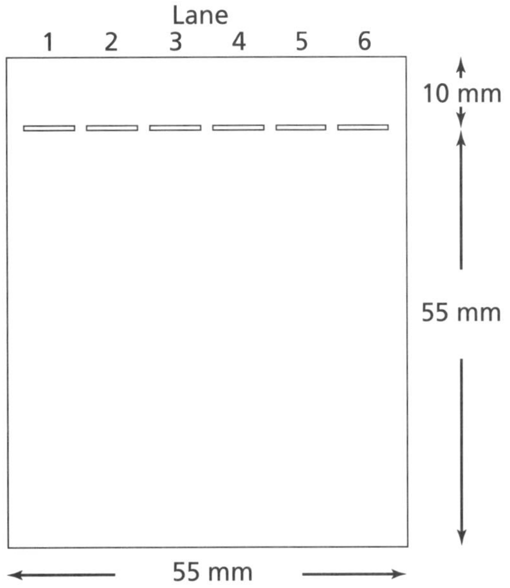

17. Compute the log of the size of each of the DNA marker sample's fragments using the data in the table on the next page. *Note: Fragment 1 is the fragment that is closest to the well.*

Table 2 Data for DNA Marker Sample (Lambda DNA/HindIII Digest)

Fragment number	Fragment length (bp)	Log	Migration distance (mm)	R_f
1				
2				
3				
4				
5				
6				
7			not imaged	———
8			not imaged	———

18. Calculate the R_f value for each fragment in the DNA marker sample (Lambda DNA/*Hin*dIII digest). Record these values in the table above.

19. Using the grid below, construct a standard curve for the DNA marker sample by plotting the R_f value of each fragment versus the log of its length.

Standard Curve for the Lambda DNA/HindIII Marker Sample

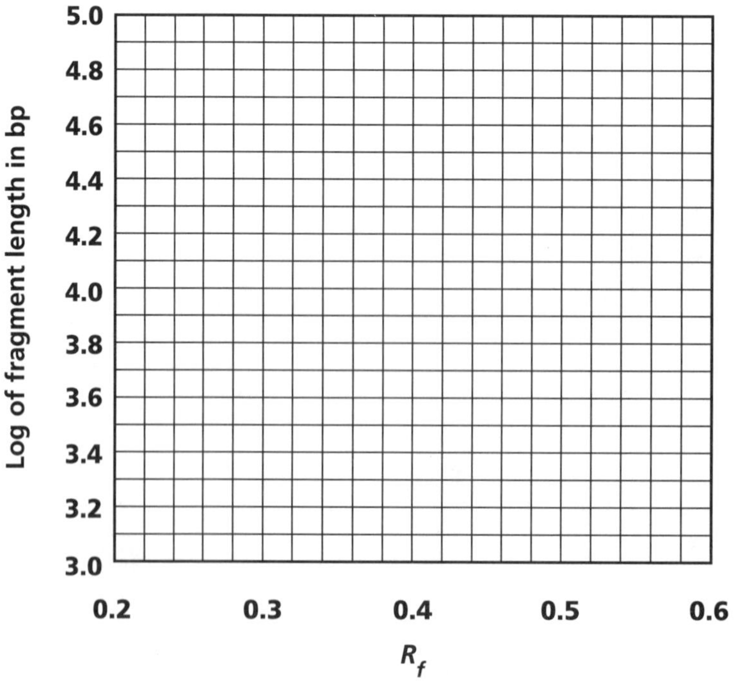

20. Calculate the R_f value for each fragment in the *Eco*RI and *Eco*RI/*Hin*dIII digests. Record these values in the data table below.

Table 3 Data for EcoRI and EcoRI/HindIII Digests

*Eco*RI					*Eco*RI/*Hind* III				
Fragment number	Migration distance (mm)	R_f	Log	Fragment length (bp)	Fragment number	Migration distance (mm)	R_f	Log	Fragment length (bp)

21. Use the standard curve you plotted in step 19 and the R_f values you calculated in step 20 to find the log of the length in base pairs (bp) of each fragment in the *Eco*RI and *Eco*RI/*Hin*dIII digests. Record the logs in the data table above. The antilog of the log for each fragment is its number of base pairs.

Analysis

22. Where on your gel are the fragments with the largest R_f values?

____

____

23. Was the DNA in any of the bands darker than in others? What could account for this?

____

____

____

____

Conclusion

24. In this lab, DNA from the same type of virus was cut with different restriction enzymes. What results would you expect from the gel electrophoresis of DNA from different organisms but cut by the same restriction enzyme?

____

____

____

____

____

____

25. Would you expect different individuals to have the same or different patterns of DNA bands? Justify your answer.

____

____

____

____

____

____

____

____

Extensions

26. Restriction endonucleases are frequently named after the genus of the organism from which the enzyme was isolated. Find out how the names are developed, and be able to identify the various parts of a name.

27. Find out how restriction enzymes are being used to develop recombinant DNA for the production of vaccines and antibiotics that are used to fight human diseases.

Name ______________________

Date ______________ Class ______________

HOLT BIOSOURCES LAB PROGRAM BIOTECHNOLOGY

D7 Laboratory Techniques: DNA Ligation

Skills

- conducting agarose gel electrophoresis

Objectives

- *Perform* a DNA ligation.
- *Use* agarose gel electrophoresis to separate DNA fragments of different sizes.
- *Analyze* DNA fragments to determine their length.

Materials

- safety goggles
- lab apron
- microtube rack
- samples in labeled microtubes:
  Tube 1 "lambda DNA/*Hind* III" (marker)
  Tube 2 "lambda DNA/*Eco*RI" (control)
  Tube 3 "lambda DNA/*Eco*RI" (ligated)
- 1–5 µL micropipets (3)
- 1× ligation buffer (2 µL)
- T4 DNA ligase (3 µL)
- watch or clock with second hand
- ice-water bath
- thermometer
- hot-water bath
- loading (tracking) dye
- 10 µL micropipetter
- micropipetter tips (3)
- agarose gel (0.8%) on gel-casting tray
- 250 mL graduated cylinder
- 1× TBE running buffer 1× (200 mL)
- 250 mL beaker
- electrophoresis system, battery-powered
- 9 V batteries (5)
- "gel handler" spatula
- DNA stain (100 mL per gel)
- staining tray
- distilled water
- metric ruler
- calculator

Purpose

You are a laboratory technician working for a company that does genetic engineering. Three batches of your DNA ligase enzymes have not been working properly. The laboratory director wants to know if the remaining supplies of enzymes are good. You know some tubes of the enzymes were left out over a hot weekend. You decide to ligate, or join, fragments of lambda DNA that have been cut with the restriction enzyme *Eco*RI. You will test your results by using agarose gel electrophoresis and then comparing the banding pattern of the ligated DNA with the same DNA prior to ligation.

Background

DNA ligation, joining fragments of DNA, is a procedure widely used for creating molecules of recombinant DNA. **Recombinant DNA** is a DNA molecule formed when fragments of DNA from two or more different organisms are joined together. Recombinant DNA is then inserted into a vector, viral or bacterial DNA, and allowed to infect target cells. The infected cells are allowed to reproduce a large number of identical cells that contain the recombinant DNA.

DNA ligases are enzymes that join fragments of DNA. These enzymes act as catalysts, joining two strands of DNA. One of the most thoroughly investigated ligases is T4 DNA ligase from the bacteriophage T4. Many other types of DNA ligases have been found in both plant and animal cells, including some mammalian cells. The construction of the first recombinant DNA molecule in the early 1970s marked the birth of the field of **genetic engineering**—moving genes from the chromosomes of one organism to those of another.

*Eco*RI is a **restriction enzyme** (RE), which is an enzyme that scans the length of a DNA molecule and cuts it only at a particular site. *Eco*RI cuts the complementary strands of a DNA molecule at a different point within the recognition sequence, resulting in a staggered cut, as shown in the diagram below.

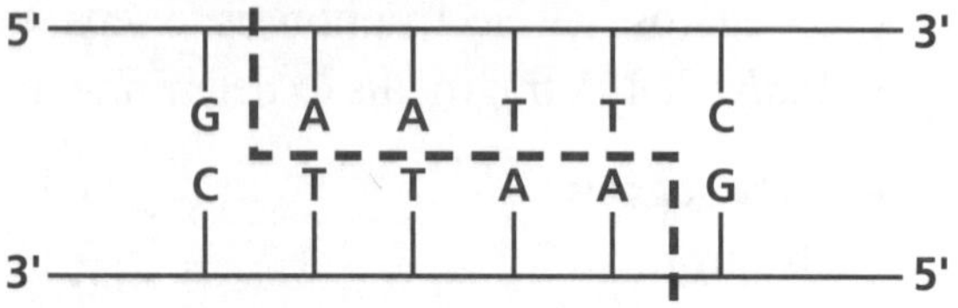

A staggered cut in a DNA molecule exposes single-stranded regions of the molecule. These single-stranded regions, called **sticky ends,** are useful in making recombinant DNA molecules. The sticky ends allow complementary regions in two DNA fragments to recognize one another and join during ligation.

Procedure

Part 1—DNA Ligation

1. Put on safety goggles and a lab apron.

2. Obtain one each of Tubes 1, 2, and 3. *Note: The sample marked "lambda DNA/HindIII" will serve as a marker, which you will use to construct a standard curve that can be used to find the length of the DNA molecules in the other two samples.* Use clean micropipets to add 2 µL of 1× ligation buffer and 3 µL of T4 DNA ligase to Tube 3, which contains 10 µL of lambda DNA that has been cut with *Eco*RI.

3. Incubate Tube 3 at 16°C for 20 minutes in an ice-water bath. At this temperature, the T4 DNA ligase catalyzes the ligation of greater than 95% of the lambda DNA fragments.

4. To inactivate the enzyme, incubate Tube 3 in a 65°C hot-water bath for 10 minutes.

5. Remove Tube 3 from the hot-water bath, and add 2 µL of loading dye. Your sample is now ready for gel electrophoresis. To test whether the ligation procedure was successful, electrophorese samples from each of Tubes 1, 2, and 3.

Part 2—Loading and Running a Gel

6. Set a micropipetter to 10 µL, and place a new tip on the end. Open Tube 1 containing lambda DNA/*Hind*III, and remove 10µL of the solution. Carefully place the solution in the well in Lane 1 of an agarose gel in a gel-casting tray. To do this, place both elbows on the lab table, lean over the gel, and slowly lower the micropipetter tip into the opening of the well before depressing the plunger. *Note: Do not jab the micropipetter tip through the bottom of the well.*

7. Using a new micropipetter tip for each tube, repeat step 6 for each of the remaining microtubes:

   Lane 2 (Tube 2) lambda DNA/*Eco*RI (control)

   Lane 3 (Tube 3) lambda DNA/*Eco*RI (ligated)

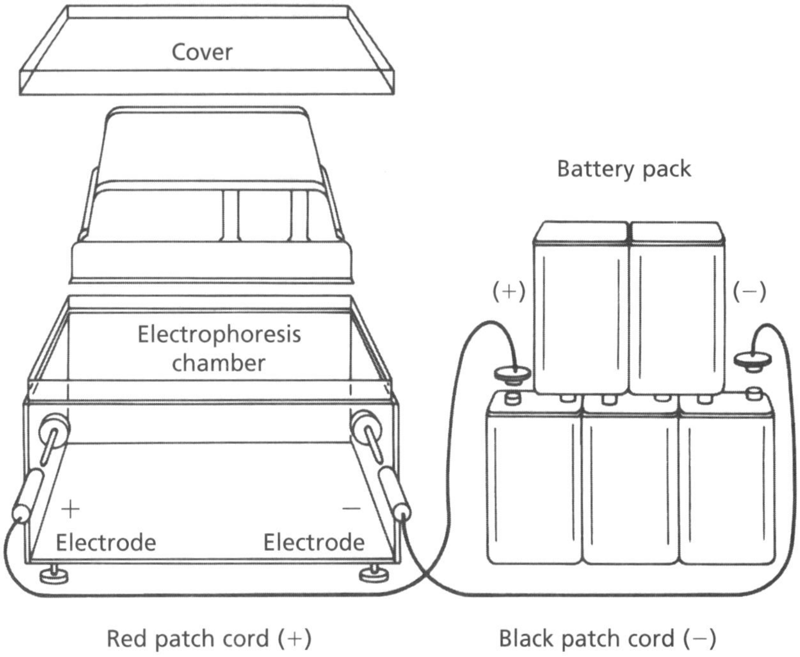

8. Carefully place your loaded agarose gel (still in the gel-casting tray) in the electrophoresis chamber of an electrophoresis apparatus, such as the one shown above. Orient the gel so that the wells are closest to the black connector, or negative electrode.

9. Slowly pour approximately 200 mL of 1× TBE running buffer into a beaker. **CAUTION: Glassware is fragile. Notify your teacher promptly of any broken glass or cuts. Do not clean up broken glass or spills unless your teacher tells you to do so. If you get a chemical on your skin or clothing, wash it off at the sink while calling to your teacher. If you get a chemical in your eyes, promptly flush it out at the eyewash station while calling to your teacher. Notify your teacher in the event of any chemical spill.** Gently and slowly pour the running buffer from the beaker into one side of the electrophoresis chamber until the gel is completely covered (1 to 2 mm above the top surface of the gel). *Note: Be careful not to overfill the chamber with buffer.*

10. Place the cover over the electrophoresis chamber. Wipe off any spills around the electrophoretic apparatus before doing the next step.

11. Connect five 9 V alkaline batteries as shown in the diagram above. **CAUTION: Do not touch both ends of the patch cords or both terminals on the battery pack at the same time.** Connect the red (positive) patch cord to both the red electrode terminal on the chamber and the red terminal on the battery pack. Follow the same procedure with the black (negative) patch cord.

12. Observe the migration of the tracking, or loading, dye down the gel toward the red (positive) electrode.

13. Disconnect the battery pack when the tracking dye has reached the end of the gel.

14. Remove the cover from the electrophoresis chamber. Lift the gel tray (containing the gel) from the cell and place the gel in the staining tray. Do this by *gently* pushing the gel off the tray using a "gel handler" spatula.

Part 3—Staining a Gel

15. To stain the gel, pour approximately 100 mL of DNA stain into the staining tray until the gel is completely covered. Do not pour stain directly onto the gel. **CAUTION: DNA stain will stain your skin and clothing. Promptly wash off spills to minimize staining.** Cover and label the staining tray. Staining will be complete in 2 to 3 hours.

16. When the gel is stained, carefully pour the remaining stain directly into a sink. Flush down the drain with water. *Note: Do not allow the gel to move against the corner of the staining tray. The gel must remain flat; if it breaks, it will be useless.*

17. To destain the gel, add enough distilled water to the staining tray to cover the gel. *Note: Do not pour water directly onto the gel. Pour water to one side of the gel.* Let the gel sit overnight (or at least 18–24 hours). After destaining the gel, the bands of DNA will appear as purple lines against a light, almost clear background.

18. Dispose of your materials according to the directions from your teacher.

19. Clean up your work area and wash your hands before leaving the lab.

Analysis

20. Use a metric ruler to measure the distance in millimeters from the well in each lane to the tracking dye and DNA bands in the same lane. *Note: Measure from the middle of a well to the middle of a band.* As you measure the distance each DNA band moved, record this measurement under "Migration distance" in the appropriate data table on the following pages. Draw the position of the DNA bands in each lane of your gel on the "gel" diagram below.

21. Compute the log of the size of each of the Lambda DNA/*Hind*III marker sample's fragments using the data in the table below. *Note: Fragment 1 is the fragment that is closest to the well.*

Data for DNA Marker Sample (Lambda DNA/HindIII Digest)

Fragment number	Fragment length	Log	Distance migrated (mm)	R_f
1	23,130			
2	9,416			
3	6,557			
4	4,361			
5	2,322			
6	2,027			
7	564			
8	125			

22. In the space below, calculate the R_f value for each of the DNA marker sample's fragments by using the following formula:

$$R_f = \frac{\text{distance the DNA fragment has migrated from the origin (gel well)}}{\text{distance from the origin to the reference point (tracking dye)}}$$

Record these values in the table above.

23. Using the grid below, construct a standard curve for the DNA marker sample by plotting the R_f value of each fragment versus the log of its length.

Standard Curve for the Lambda DNA/Hind*III *Marker Sample

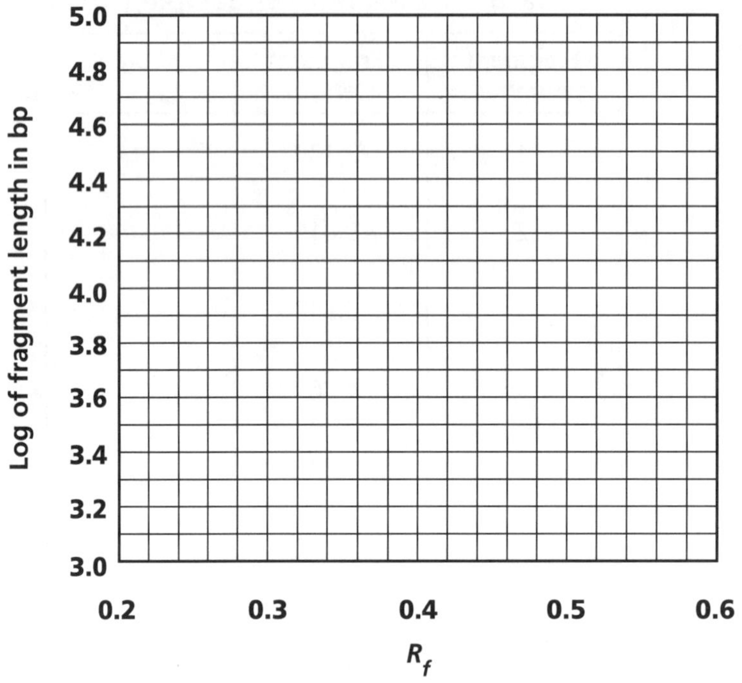

24. In the space below, calculate the R_f value for each fragment in the *Eco*RI (control) and *Eco*RI (ligated) samples. Record these values in the data table on the next page.

Data for EcoRI (Control) and EcoRI (Ligated)

*Eco*RI (control)					*Eco*RI (ligated)				
Fragment number	Migration distance (mm)	R_f	Log	Fragment length (bp)	Fragment number	Migration distance (mm)	R_f	Log	Fragment length (bp)

25. Use the standard curve you plotted in step 23 and the R_f values you calculated in step 24 to find the log of the length in base pairs (bp) of each fragment in the *Eco*RI (control) and *Eco*RI (ligated) samples. Record the logs in the data table above. The antilog of the log for each fragment is its number of base pairs.

Conclusions

26. Based on your gel analysis, was your ligation reaction successful? Explain your answer.

____

____

____

____

____

____

27. What is your report concerning the condition of your company's supply of ligation enzyme?

____

____

28. Compare the number of base pairs in the bands in Lane 2 with the number of base pairs in the band in Lane 3. What conclusion can you draw from this comparison?

____

____

____

____

____

____

Extensions

28. Find out the importance of the Ti plasmid in the formation of recombinant DNA in plants.

29. When genetic engineering became a reality in the early 1970s, biologists were concerned about the potential dangers of producing new strains of pathogenic and lethal microorganisms. Find out what strict protocols are in place today to prevent such misuse of recombinant DNA technology.

Name ______________________

Date ____________ Class ____________

D8 Experimental Design: Comparing DNA Samples

Prerequisites

- **Biotechnology D5—Laboratory Techniques: Introduction to Gel Electrophoresis** on pages 19–24
- **Biotechnology D6—Laboratory Techniques: DNA Fragment Analysis** on pages 25–32

Review

- use of restriction enzymes to cut DNA
- procedure for analyzing a gel

Williams & Associates

Hollywood, California

December 10, 1997

Caitlin Noonan
Research and Development Division
BioLogical Resources, Inc.
101 Jonas Salk Dr.
Oakwood, MO 65432-1101

Dear Ms. Noonan,

As a lawyer here in Hollywood, I represent many actors and actresses. Following the recent death of actress Jessica Coleman, I became the executor of Ms. Coleman's will. A few days later, a woman by the name of Ms. Wilson arrived at my office claiming to be Ms. Coleman's long-lost identical twin. I read through the will, but found no reference specific to any individual. Instead, the will simply states that the estate is to be divided equally among Ms. Coleman's closest living relatives. Until now, it was presumed that the entire estate would go to her only daughter. However, Ms. Wilson has provided a great deal of evidence to support her claim, and blood tests have not ruled out the possibility that the woman could be Ms. Coleman's twin.

We have blood samples from Ms. Coleman and Ms. Wilson. We are requesting that you perform the necessary DNA tests to determine whether Ms. Wilson could be the twin of the deceased. Please contact me as soon as possible.

Sincerely,

Matthew Williams

Matthew Williams
Attorney at Law
Williams & Associates

BioLogical Resources, Inc. Oakwood, MO 65432-1101

M E M O R A N D U M

To: Team Leader, Genetics Dept.

From: Caitlin Noonan, Director of Research and Development

The case described in the attached letter sounds like a challenging one. Our research teams will be using the DNA fingerprinting procedure to compare the DNA of Ms. Coleman and Ms. Wilson. I have had the Biochemistry department extract DNA from each of the blood samples supplied by Mr. Williams, and they have also cut the DNA samples using restriction enzymes.

I want your research teams to separate the DNA fragments that resulted when restriction enzymes cut the DNA. These samples, which are ready for the gel electrophoresis procedure except for the addition of loading dye, have been delivered to your lab. You will also be provided with a marker sample containing RFLPs of known lengths. The lengths of these fragments, from fragment 1 to fragment 8, are as follows: 23,130; 9,416; 6,557; 4,361; 2,322; 2,027; 564; and 125. The logs of these fragments are 4.364, 3.974, 3.817, 3.640, 3.366, 3.307, 2.751, 2.097, respectively. Use the marker sample to determine the lengths of the RFLPs in the test samples. After your team has completed the gel electrophoresis and DNA fragment analysis, the Biochemistry team will complete the DNA fingerprints for each test sample.

Please be very thorough in recording your procedure. This may become important if the case ever develops into a law suit and we are asked to explain our procedure to a jury.

Proposal Checklist

Before you start your work, you must submit a proposal for my approval. **Your proposal must include the following:**

______ • the **question** you seek to answer

______ • the **procedure** you will use

______ • detailed **data tables** for recording measurements and calculations

______ • a complete, itemized list of proposed **materials** and **costs** (including use of facilities, labor, and amounts needed)

Proposal Approval: ______________________________

(Supervisor's signature)

Report Procedures

When you finish your analysis, prepare a report in the form of a business letter to Mr. Williams. **Your report must include the following:**

______ • a paragraph describing the **procedure** you followed to test the DNA from each subject

______ • a complete **data table** showing the R_f values for each sample's RFLPs

______ • a **graph** plotting the logs of the fragment lengths against R_f values for the marker sample

______ • your **conclusions** about whether the subjects may be related

______ • a detailed **invoice** showing all materials, labor, and the total amount due

Safety Precautions

- Wear safety goggles and a lab apron.
- Glassware is fragile. Notify your teacher promptly of any broken glass or cuts. Do not clean up broken glass or spills unless your teacher tells you to do so.
- Never use electrical equipment around water or with wet hands or clothing. Never use equipment with frayed cords.
- Wash your hands before leaving the laboratory.

Disposal Methods

- Dispose of waste materials according to instructions from your teacher.
- Place used paper towels and plastic zippered bags in a trash can.
- Place broken glass, microtubes with DNA samples, gel, unused 1× TBE buffer, unused loading dye, unused DNA stain, and disinfectant solution in the separate containers provided.
- Wash reusable materials such as glassware and lab utensils, and return them to the supply area.

FILE: Williams & Associates

MATERIALS AND COSTS (Select only what you will need. No refunds.)

I. Facilities and Equipment Use

Item	Rate	Number	Total
facilities	$480.00/day		
personal protective equipment	$10.00/day		
clock or watch with second hand	$10.00/day		
gel-casting tray (with gel in plastic bag)	$5.00/day		
6-well gel comb	$2.00/day		
microtube rack	$5.00/day		
10 μL fixed-volume micropipetter	$2.00/day		
gel spatula	$3.00/day		
staining tray	$5.00/day		
gel electrophoresis system	$10.00/day		
freezer	$10.00/day		
metric ruler	$1.00/day		
calculator	$5.00/day		

II. Labor and Consumables

Item	Rate	Number	Total
labor	$40.00/hour		
graph paper	$0.25/sheet		
microtubes with cut DNA samples	$10.00 each		
1–5 μL micropipets	$2.00 each		
5–200 μL micropipetter tips	$2.00 each		
1× TBE buffer	$0.20/mL		
loading dye	$0.10/μL		
DNA stain	$0.10/mL		
distilled water	$0.10/mL		
disinfectant solution	$2.00/bottle		
paper towels	$0.10 each		
nontoxic, permanent marker	$2.00 each		

Fines

Item	Rate	Number	Total
OSHA safety violation	$2,000.00/incident		
		Subtotal	
		Profit Margin	
		Total Amount Due	

Name ______________________________

Date ______________ Class ______________

HOLT BIOSOURCES
LAB PROGRAM
BIOTECHNOLOGY

D9 Laboratory Techniques: Introduction to Fermentation Technology

Skills

- using aseptic technique
- culturing bacteria

Objectives

- *Assemble* a working fermenter.
- *Produce* the antibiotic bacitracin using *Bacillus licheniformis* in the fermenter.
- *Perform* a qualitative test of the antibiotic produced by *Bacillus licheniformis.*

Materials

- safety goggles, gloves, and a lab apron
- petroleum jelly
- 3-hole rubber stopper, #8
- rigid plastic tubing, 11 in. length, 3/16 in. outside diameter (2)
- rigid plastic tubing, 5 in. length, 3/16 in. outside diameter
- vinyl air-line tubing, 3 in. length, 5/16 in. outside diameter (3)
- 1 L Erlenmeyer flask
- 70% ethanol or isopropanol (rubbing alcohol)
- in-line filters (2)
- vinyl air-line tubing, 3 ft length, 5/16 in. outside diameter
- screw compress clamp
- agar slant of *Bacillus licheniformis*
- sterile inoculating loops (2)
- 10 mL graduated cylinder
- 100 mL bottle of sterile NYSM inoculum medium with a magnetic stir bar
- magnetic stirrer
- 500 mL bottle of sterile NYSM fermentation medium
- air pump
- 10 mL syringe without a needle
- test tube, 13 × 100 mm
- sterile pipet
- pH paper or pH meter
- petri dish of *Bacillus thuringiensis* var. israelensis (2)
- forceps
- sterile 5 × 50 mm pieces of filter paper for antibiotic assay (12)

Purpose

You have just been hired as an industrial microbiologist for a large pharmaceutical company. The company is a major supplier of the antibiotic bacitracin. Your job will be to monitor the production of bacitracin and to conduct research to improve the fermentation technology by which it is produced. To learn about the production of bacitracin through fermentation, first you will observe the process using a fermenter made from simple lab equipment. Then, you will assay, or test, the product you made for its ability to kill bacteria.

Background

The term *fermentation* has two meanings. In cell metabolism, the term *fermentation* refers to a metabolic process that follows glycolysis (the splitting of glucose molecules) in the absence of oxygen. Although fermentation occurs in all types of cells, the most familiar example of fermentation is the production of carbon dioxide and ethyl alcohol by microorganisms such as bacteria and fungi. The breakdown of organic compounds during cell metabolism also enables cells to obtain the energy and raw materials they need to make other organic compounds. The term *fermentation* is also used in the field of **biotechnology,** the application of biological techniques to industry. In biotechnology, **fermentation** is the controlled production of commercially useful products by the action of microorganisms. These products are produced in containers called **fermenters.** Products resulting from fermentation include beer, wine, cheese, citric acid, solvents, insecticides, fragrances, and vitamins, as well as a host of antibiotics, enzymes, and other organic compounds with substantial commercial value.

Recombinant DNA technology and the science of genetic engineering have increased the number of commercially important chemicals that can be produced by fermentation technology. The genes that code for the production of a number of substances have been identified and introduced into well-understood organisms such as *E. coli.* These organisms can be economically mass-produced to increase the yield of desirable products. Today, virtually all antibiotics are produced by isolating and purifying products made by organisms grown in fermenters. For example, bacitracin, which was discovered at almost the same time as penicillin, is produced through fermentation by the bacterium *Bacillus licheniformis.* **Bacitracin** is a polypeptide antibiotic that acts as a cell wall inhibitor, particularly of gram-positive bacteria.

Part 1—Fermentation Vessel Assembly

1.  Put on safety goggles, gloves, and a lab apron.

2. Apply a small amount of petroleum jelly to each hole of a #8, 3-hole rubber stopper. Carefully insert three pieces of rigid plastic tubing (two 11 in. pieces and one 5 in. piece) through the holes of the rubber stopper, as shown in the diagram at right. The shorter rigid tubing will be the air outlet. One of the two longer tubes will be the air inlet, and the other will be a sampling port.

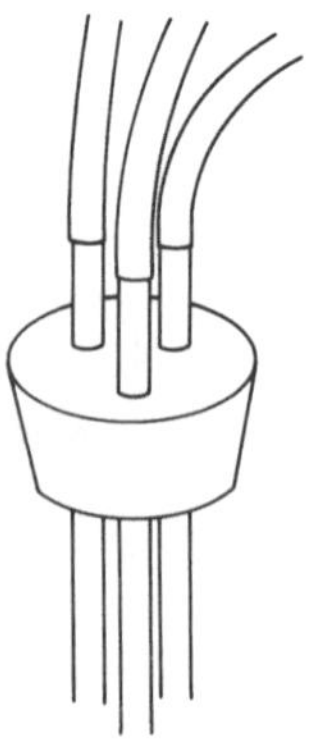

3. Attach a 3 in. piece of vinyl tubing to the top end of each piece of rigid tubing, as shown in the diagram above.

4.  Soak the top (stopper and tubing) assembly you made in steps 2 and 3 in 70% ethanol for about five minutes to sterilize it. Prior to use, remove the assembly from the alcohol, hold it over a sink, and quickly shake off any excess alcohol still present on the assembly and in the tubes. **CAUTION: Be careful not to get any alcohol on you or your classmates.**

5. Immediately place the top assembly onto a flask, making sure that the rubber stopper seals the opening of the flask completely.

6. Attach an in-line filter to each end of the air-inlet and air-outlet tubings, and then attach a 3 ft piece of vinyl air-line tubing to the filter on the air-inlet tubing. Finally, attach a screw compress clamp to the sampling port, as shown in the diagram below. Your fermenter is now ready for use.

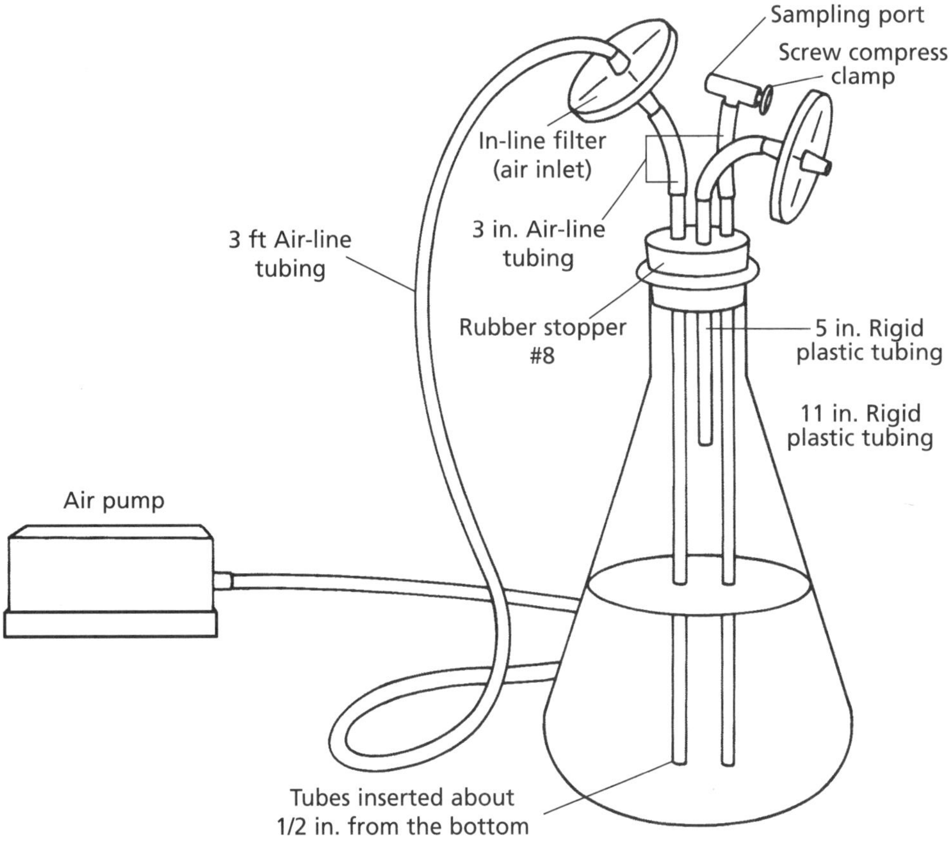

Part 2—Preparing Inoculum and Inoculating the Fermenter

Before beginning the process of fermentation, an inoculum, which is a quantity of bacteria, must be "built up." The inoculum buildup has the following three purposes: to maximize the number of cells; to synchronize their growth; and to prepare cells that are in the proper growth stage. The life cycle of a colony of bacteria can be divided into the following three stages: vegetative growth, sporulation (spore formation), and death. Some of the useful products bacteria produce through fermentation are made during the vegetative stage, while others, such as bacitracin, are made during the sporulation stage.

7. 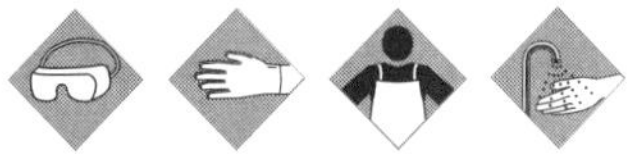 Put on safety goggles, gloves, and a lab apron. *Note: Although these microbes are naturally occurring and nonpathogenic, you must use personal protective equipment and aseptic technique throughout the procedure to minimize the risk of any contamination.*

8. Measure 1 mL of sterile NYSM inoculum medium, and add it to an agar slant of *B. licheniformis.*

9. Using a sterile inoculating loop, carefully scrape off the growth from the slant into the liquid. Pour the liquid containing the bacteria back into the 100 mL NYSM inoculum bottle.

10. Make sure the cap of the 100 mL NYSM inoculum bottle is screwed on tightly, and then place the bottle on its side on a magnetic stirrer.

11. Incubate the inoculum bottle at room temperature or in a 30°C incubator, if available. When growth becomes apparent (usually within 18–24 hours), the inoculum is ready and can be used to inoculate (add bacteria to) your fermenter.

12. Working quickly, partly remove the top assembly from the flask of your fermenter by pulling up on one of the rigid tubings with one hand. With your other hand, pour the contents of a 500 mL bottle of NYSM fermentation medium into the flask.

13. Attach the air-inlet tubing to an air pump, and adjust the air flow so that air bubbles gently into the medium. This will provide oxygen to cells as well as provide a positive pressure to prevent contamination.

14. Using the technique described in step 12, carefully add the inoculum you made in steps 8–11 to your fermenter. Allow the fermentation process to run for 48 hours.

Part 3—Sampling and Testing Procedures

To monitor the progress of your fermentation run, you will need to periodically withdraw small samples of your fermentation broth (inoculated fermentation medium) and test the samples. Test the pH of your broth at the start of the fermentation run, at least once during the vegetative phase (at about 24 hours), and again when the broth is harvested (at about 48 hours). Assay (test) for bacitracin production after 24 hours and again at the end of 48 hours.

Almost all *Bacillus* bacteria, when grown in a liquid medium, show a distinctive change in pH during their growth cycle. An example of this change is shown in the graph below.

Growth and pH Curves for *Bacillus* in a Fermenter

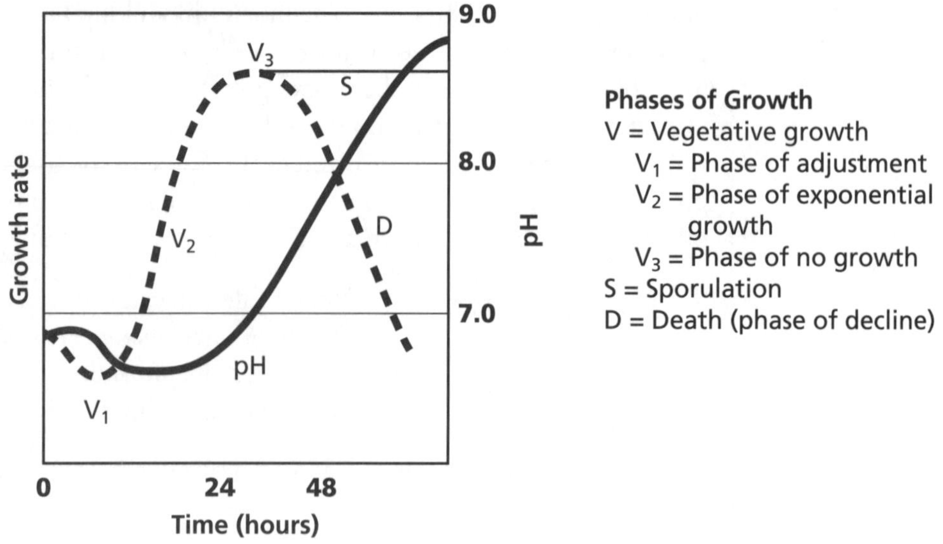

15. To remove a sample of fermentation broth, attach a syringe (without a needle) to the sampling port of your fermenter. Loosen the screw compress clamp, and pull on the syringe handle to withdraw 1–2 mL of broth.

16. Expel the sample into a small test tube, and label it accordingly.

17. Tighten the screw compress clamp, and rinse the end of the sampling port with a small amount of either 70% ethanol or rubbing alcohol to prevent the contamination of your fermenter.

18. Using a sterile pipet, place a drop of the broth culture on a piece of pH paper. Record the pH value in the data table below.

19. Repeat steps 15–18 at the time periods shown in the table below.

Results of pH Test and Bacitracin Assay

Variable	Incubation time		
	Start	24 hours	48 hours
pH			
Bacitracin assay (size of zone of inhibition) (mm)	———		

20. Use the remainder of the samples removed after 24 and 48 hours to assay for bacitracin. Using forceps, dip a sterile strip of filter paper into the sample. Then carefully place the strip onto the agar in the center of a petri dish previously inoculated with *B. thuringiensis*, as shown in the diagram below.

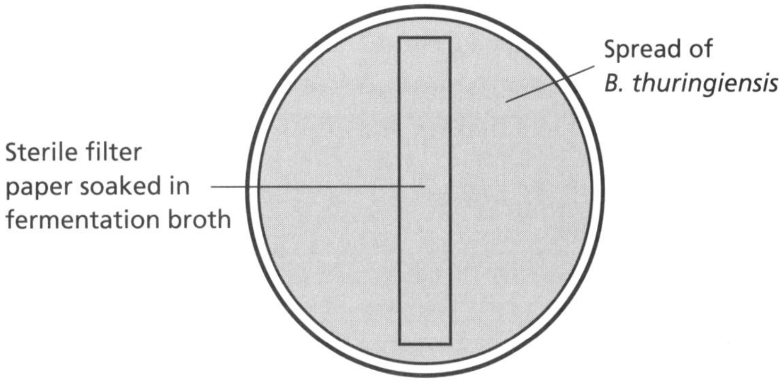

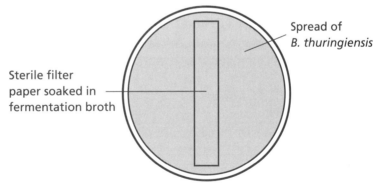

21. Incubate each petri dish until growth is visible (usually within 48 hours). At that point, you should see a zone of inhibition (a clear zone of no growth) if bacitracin is present.

22. Dispose of your materials according to the directions from your teacher.

23. Clean up your work area and wash your hands before leaving the lab.

Analysis

24. What was the purpose of the in-line filters in your fermenter?

25. What happened to the pH of the fermentation broth during the fermentation run? How do you account for this change?

Conclusions

26. Do you think that the production of antibiotics is important to the survival of bacteria in nature? Justify your answer.

27. Bacteria are used to produce compounds that are found only in certain species, such as human insulin. How do you think this is accomplished?

Extensions

28. Search an on-line data base or do library research to learn about how recombinant DNA and industrial fermentation techniques are used in the production of other economically important chemicals.

29. *Microbiologists* study microorganisms, such as bacteria and protists. Find out about the training and skills required to become a microbiologist, and determine what types of industries employ microbiologists.

Name ______________________

Date ____________ Class ____________

HOLT BIOSOURCES
LAB PROGRAM
BIOTECHNOLOGY

D10 Laboratory Techniques: Ice-Nucleating Bacteria

Skills

- making observations
- graphing data

Objectives

- *Observe* the effects of ice-nucleating proteins on ice formation, the freezing temperature of water, and the heat of crystallization.
- *Calculate* the cooling rate of water.

Materials

- safety goggles
- lab apron
- wax pencil or felt-tip marking pen
- test tubes (3)
- test-tube rack
- distilled water (15 mL)
- ice-nucleating protein granules
- rubber stopper
- microplate
- aluminum foil (4 in. × 6 in.)
- graduated plastic pipets (4)
- freezer
- stopwatch or clock with second hand
- cardboard strips, 1/2 in. × 2 in. (2)
- −30°C to 50°C thermometers (2)
- stapler
- calculator

Purpose

You are a plant physiologist working for the state department of agriculture. You live in a state that grows a majority of the country's oranges and lemons. Once oranges and lemons form on the trees, they are very susceptible to damage by freezing temperatures. One way growers protect their fruit from freezing temperatures is to spray the trees with water. The resulting layer of ice protects the fruit from further damage. You have read about a kind of bacteria, called ice-plus bacteria, that causes water to freeze at a higher temperature. You wonder if the oranges and lemons would have less damage if the protective coating of ice formed at a higher temperature. You decide to run a series of tests to find out at what temperature the ice-nucleating protein produced by ice-plus bacteria causes water to freeze.

Background

A great deal of research has been done lately concerning the freezing temperature of water. The common misconception is that water freezes at 0°C. In fact, freezing rarely starts at 0°C. Water in its purest state can be "supercooled" to as low as −40°C without ice formation.

An "ice nucleator" helps water to freeze by attracting the water molecules and slowing them down. A **nucleator** is any foreign particle in water that allows the freezing process to begin. The ice-nucleating protein (INP) that will be used in this investigation is derived from the naturally occurring bacterium *Pseudomonas syringae.* This form of *Pseudomonas* is sometimes called the "ice-plus" variety because it contains a gene that promotes the formation of proteins that serve as nucleators.

Another feature of an ice-nucleating protein is that it is capable of initiating the freezing process at a higher temperature. The result is that water treated with an ice-nucleating protein freezes faster, more completely, and over a wider range of conditions.

The ability to have ice form at higher temperatures has many commercial applications, such as weather modification (including snow making), natural cooling of water as a refrigeration source, water purification, and construction in the Arctic and Antarctic. On the other hand, the formation of ice causes wide-scale damage to some crops due to the presence of ice-plus bacteria.

However, another strain of *Pseudomonas*, an "ice-minus" strain, also occurs naturally. It is identical to the ice-plus variety except that it lacks the ice-promoting gene. Ice-minus bacteria do not freeze until the temperature goes below −7°C. Since the formation of ice causes frost damage, the absence of the ice-nucleating protein protects some plants from frost as much as seven degrees below the normal freezing point of water. When these ice-minus bacteria are applied to a plant, water can be supercooled rather than freeze and damage the plant. The economic incentive of protecting plants and crops from frost damage is enormous.

Procedure

Part 1—Observing the Effects of Ice-Nucleating Protein

1. Put on safety goggles and a lab apron.

2. Use a wax pencil or felt-tip marking pen to label a test tube "INP-treated water." **CAUTION: Glassware is fragile. Notify your teacher promptly of any broken glass or cuts. Do not clean up broken glass or spills unless your teacher tells you to do so.** Add 10 mL of distilled water to the test tube. Then add 3 to 4 granules of the ice-nucleating protein to the test tube. Put a rubber stopper in the mouth of the test tube. Mix the contents well by inverting the test tube several times.

3. With the wax pencil, write your name on the bottom of a microplate. Wrap the top of the microplate with the piece of aluminum foil. Make an imprint of the wells on the foil by gently rubbing the foil with your hand.

4. Using a graduated plastic pipet, place one drop of the INP-treated water into 30 of the wells formed by the foil. *Note: For the best results, keep the water droplets as small as possible.* Save the remaining INP-treated water for later use.

5. With a clean, graduated plastic pipet, place one small drop of distilled water into 30 adjacent wells formed by the foil. *Note: Again, keep the water droplets as small as possible.*

6. Place your microplate in a freezer. Make observations of your plate every three minutes for about 15 minutes or until one set of droplets freezes. Watch for any changes in freezing between the plain distilled water droplets and the INP-treated water droplets. *Note: Keep the freezer door open only long enough to make your observations.*

- Did the INP-treated water droplets freeze faster than the plain water droplets? Explain your answer.

Part 2—Measuring the Effects of INP on Freezing Temperature, Cooling Rate, and Heat of Crystallization

7. Use a wax pencil to label two test tubes *Tube 1* and *Tube 2*. Add your initials to each tube.

8. Use a clean, graduated plastic pipet to transfer 3 mL of the INP-treated water to Tube 1. Place the tube in a test-tube rack.

9. Using a clean, graduated plastic pipet, add 3 mL of distilled water to Tube 2, and place the tube in the rack next to Tube 1.

10. Fold two cardboard strips in half lengthwise. Place the fold of one piece of cardboard over the top of one of the thermometers. Staple each end to hold the cardboard securely to the thermometer as shown in the diagram below. *Note: Be careful not to break the thermometer with the stapler.* Repeat for the second thermometer.

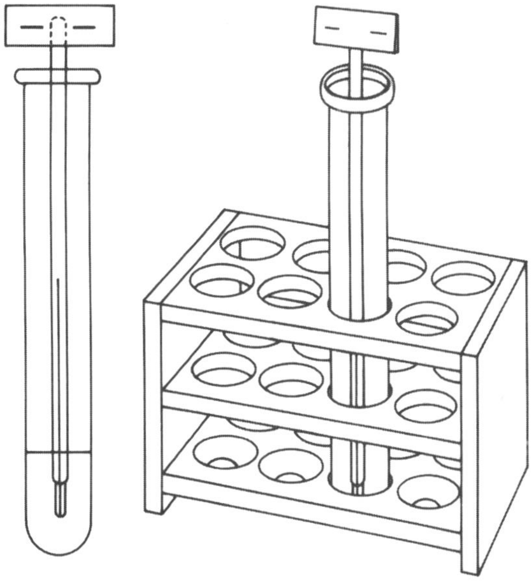

11. Insert a thermometer into each test tube. Use the cardboard strips to position each thermometer so that it does not touch the glass. Put the test tubes and test-tube rack in the freezer.

12. Take temperature readings of Tube 1 and Tube 2 at five-minute intervals for a minimum of 50 minutes. Do not touch the thermometers or tubes. Make careful observations of any ice formation during this period. *Note: Keep the freezer door open only long enough to take measurements.* Record the temperature readings and observations in the tables on the next page.

Tube 1 INP-Treated Water

Time (min.)	Temp. Tube 1 (°C)	Cooling Rate (°C/min)	Observations
0			
5			
10			
15			
20			
25			
30			
35			
40			
45			
50			
55			
60			

Tube 2 Distilled Water

Time (min.)	Temp. Tube 2 (°C)	Cooling Rate (°C/min)	Observations
0			
5			
10			
15			
20			
25			
30			
35			
40			
45			
50			
55			
60			

13. Calculate the rate of cooling for each test tube using the following equation. Record your findings in the appropriate data table on the previous page.

$$\frac{°C}{min} = \frac{T_2 - T_1}{t_2 - t_1}$$

where T_1 = temperature at time interval *start* t_1 = time interval *start*
T_2 = temperature at time interval *stop* t_2 = time interval *stop*

14. On the grid below, graph temperature versus time during freezing for Tube 1 and Tube 2. Using the graph, determine the heat of crystallization. The heat of crystallization is the amount of heat released when a liquid is transformed into ice crystals. The heat of crystallization for this lab is found by finding the lowest temperature at which water starts to freeze and the highest temperature at which freezing is completed. Find the difference between the two temperatures. Multiply that number by 18 to determine the heat of crystallization.

Change in Temperature During Freezing

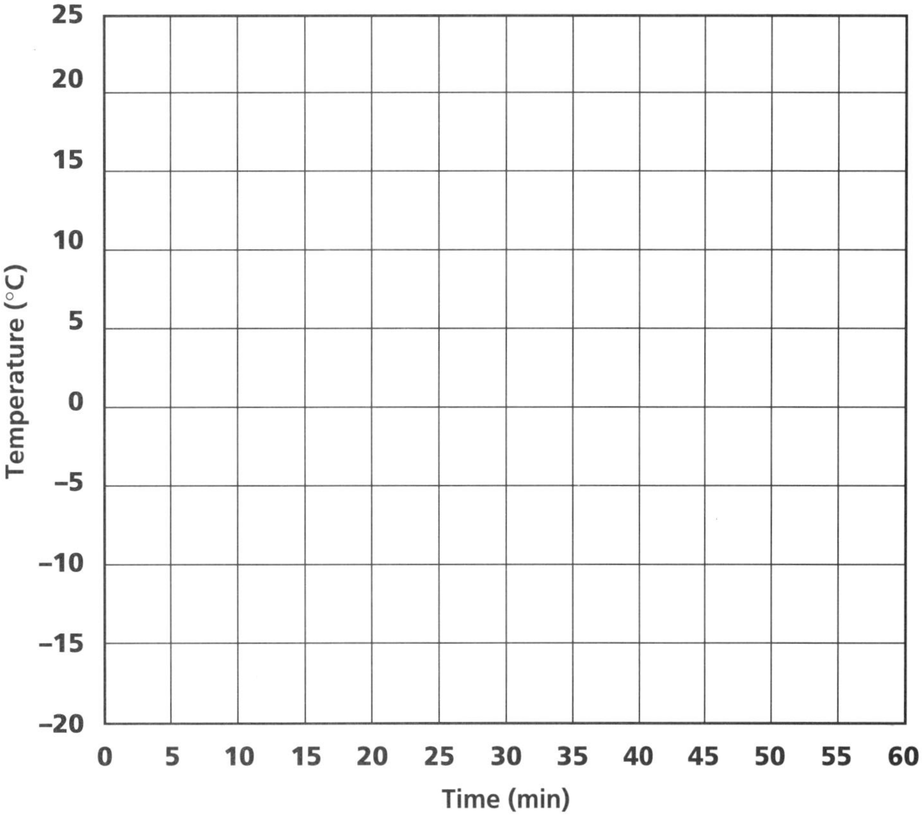

15.  Dispose of your materials according to the directions from your teacher.

16.  Clean up your work area and wash your hands before leaving the lab.

Analysis

17. Which of the two test tubes from step 11 first started to display signs of ice crystal formation?

____

____

18. At what temperature did the distilled water and the INP-treated water begin to freeze? Were these temperatures what you expected them to be?

19. Explain the differences in the temperatures for the first signs of freezing.

Conclusions

20. According to your data and graph, what effect did the INP have on the formation of ice? the cooling rate of water?

21. What other substances, organic or inorganic, might be used as ice nucleators?

22. Based on the results of your experiments, what would be the next step in determining if oranges and lemons would be protected by the addition of ice-plus bacteria?

Extensions

23. Find out how the application of ice nucleators can be used to modify weather, specifically how they can be used to cause rain or snow to fall.

24. Find out how ice-nucleating proteins are used in nature by animals such as insects, amphibians, and reptiles to help them survive freezing during winter.

Name ______________________

Date ____________ Class ____________

HOLT BIOSOURCES LAB PROGRAM BIOTECHNOLOGY

D11 Laboratory Techniques: Oil-Degrading Microbes

Skills

- practicing aseptic technique
- developing a dynamic model of an oil spill
- controlling variables in an experiment
- collecting, organizing, and graphing data

Objectives

- *Compare* the physical characteristics of oil before and after the action of oil-degrading microbes.
- *Identify* which microorganisms are useful in cleaning up an oil spill.

Materials

- safety goggles
- gloves
- lab apron
- disinfectant solution in squeeze bottle
- paper towels
- plastic jars with lids (3)
- wax pencil
- distilled water (90 mL)
- refined oil in dropper bottle
- 1.5 g of nutrient fertilizer
- scoop
- disposable pipets (2)
- *Pseudomonas* culture
- *Penicillium* culture
- density indicator strips (3)
- incubator
- biohazard waste disposal container

Purpose

You are a marine ecologist who works for a large petroleum company. One of the problems you must solve is how to clean up the environment following an oil spill. You believe that some of the present, mechanical methods of cleaning up oil are not effective. You want to explore new methods. You have heard that certain kinds of microbes can be used to digest the spilled oil. Before an oil spill happens, you want to find out if microbes really can digest oil. You set up a controlled experiment to find out which of two different microbes digests oil most efficiently.

Background

Through the media, the public has been made increasingly aware of the hazards to the environment caused by oil spills. A single gallon of oil can spread thinly enough to cover four acres of water. Some of the oil evaporates; some is broken down by radiant energy; and some emulsifies, or breaks down into small pieces, to form a heavy material that eventually sinks to the bottom of the ocean. This heavy material endangers birds, marine mammals, and other forms of sea life.

Oil spills can be cleaned up using mechanical devices such as skimmers and barriers. Other cleanup methods include the use of chemical dispersants and solvents and the burning of oil. However, some microorganisms can break down the various hydrocarbons in an oil spill. They may be the best, most environmentally safe prospect for cleaning up oil spills. These "oil hungry" microbes convert oil into food for themselves, rendering it nontoxic and allowing it to be assimilated safely into aquatic food webs. The use of living organisms to repair environmental damage is known as **bioremediation.**

Procedure

Part 1—Inoculating Oil With Microorganisms

1. Put on safety goggles, gloves, and a lab apron.

2. Use disinfectant solution to sterilize the top of your lab table. To do this, spread disinfectant solution over the entire work area. Wipe the area clean with paper towels. Dispose of the paper towels as indicated by your teacher.

   ♦ Why is it important to sterilize your work area before you begin this investigation?

   ____

   ____

3. Use a wax pencil to label one plastic jar *Control*, a second plastic jar *Pseudomonas*, and a third plastic jar *Penicillium*. Also write the date and your name and class period on each jar.

4. Pour about 30 mL of distilled water into each jar so that each jar is about half full.

5. Add about 20 drops of refined oil to form a thin layer in each jar.

6. Using a scoop, add a pinch of fertilizer to the water and oil mixture. This fertilizer is the kind used in real oil spills and will coagulate the oil and provide nutrients for microbial growth.

7. Using disposable pipets, inoculate the appropriate jars with 5 mL (two droppers full) of a *Pseudomonas* culture and 5 mL of a *Penicillium* culture. The control jar receives no microorganisms. **CAUTION: Always practice aseptic technique when using bacteria in the lab.**

8. Secure the top on each jar, and invert it several times to mix the oil with the jar's other contents. Similar wave action occurs in the ocean and increases the amount of dissolved oxygen in the water.

   ♦ Why is it desirable to increase the amount of dissolved oxygen in the water?

   ____

   ____

9. Number the bars on three density indicator strips from 1 to 5, with 5 for the darkest. Place a density strip on each jar so that the tops of the bars are below the water level. This density strip will allow you to quantify the amount of microbial growth in each jar. As the microbes grow, the water will become cloudy, or turbid. As the microbe population increases, you will be able to see fewer bars through the water. Hold up the jar with the density strip opposite you. Place a piece of white paper behind the jar to provide a background. Look through the water in the jar to determine the number of density bars that have disappeared.

10. Record your observations for today, day 0, in the data table below.

Effects of Microorganisms on Oil

Day	Organism	General appearance of oil	Color of oil	Turbidity of water (number of bars that disappear)
0				
1				
2				
3				
4				

11. Incubate the jars with caps half loosened at 30°C. If an incubator is not available, place the jars in a warm spot in the classroom.

♦ Why is it necessary to incubate the jars?

____

____

____

____

♦ Why is it necessary to leave the jar caps half loosened during incubation?

____

____

12. Dispose of your materials according to the directions from your teacher.

13. Sterilize your work area as described in step 2. Wash your hands with antibacterial soap before leaving the lab.

Part 2—Observing the Effects of Microbes on Oil

14. Observe your tubes every 24 hours for 4 days. Look for any signs of oil degradation, such as change in color, the formation of tiny oil droplets, break-up of the oil layer into smaller fragments, or changes in texture. Microbial growth will be observed mainly at the interface (boundary) between the oil and water. Record your observations of the general appearance of the degrading oil, its color, and the turbidity of the water in the data table on the previous page. Each day after you make your observations, temporarily tighten each jar lid and invert each jar once or twice to increase the dissolved oxygen content in the water and to further mix the oil with the jar's other ingredients.

♦ Which microorganisms do you think will degrade the most oil?

____

____

Analysis

15. Using the grid below, prepare a graph showing the change in turbidity that occurs in each jar. The turbidity (number of bars that disappear) should be shown on the y-axis, with time plotted on the x-axis. Use different colored pencils to draw the curve for growth of both organisms and the control on the same graph. Be sure to make a legend that indicates which color pencil represents which organism and the control.

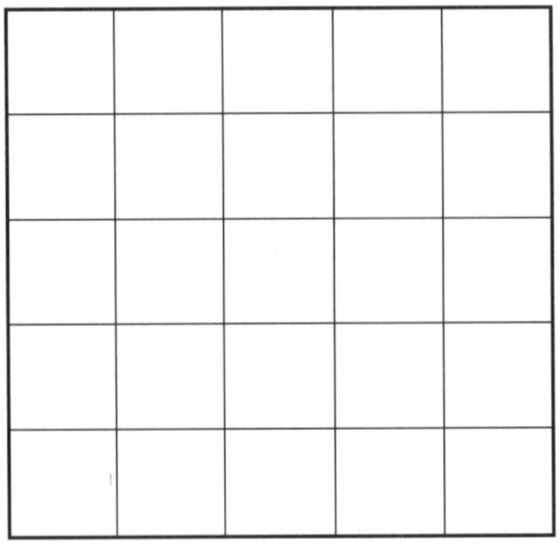

16. Describe any changes in the physical characteristics and appearance of the oil on day 1 and beyond. Discuss the possible causes for such changes.

17. What does an increase in turbidity indicate?

18. What is the purpose of the control?

19. What is the purpose of starting with distilled water rather than tap or pond water?

20. How does the procedure you used in this lab differ from an actual oil-spill cleanup?

Conclusions

21. What would happen to the growth of the microbes if no fertilizer were added?

22. Which microbe degraded the oil better?

23. What advantages, if any, can you think of for using a mixture of microorganisms rather than just one kind of microorganism to degrade the oil?

24. An environmentalist may argue that as damaging as an oil spill can be to the environment, adding fertilizer and microbes is worse. Explain this argument.

25. If you were an oil-company executive and had to decide how to clean up an oil spill from a tanker, what would you recommend? Explain your methods.

Extensions

26. *Marine biologists* study the relationships of living things and their environment in Earth's seas and oceans. Some marine biologists study the populations and communities that make up a specific ecosystem. Others study the effects of pollution and climate changes on the environment as a whole. Find out about the training and skills required to become a marine biologist.

27. Do library research or search the Internet to investigate how effective bioremediation has been in the cleanup of oil spills.

28. Conduct this experiment again without adding any fertilizer to the jars. Compare the results of this experiment with the results of the experiment in which fertilizer was added along with the oil.

Name ______________________________

Date ______________ Class ______________

D12 Experimental Design: Can Oil-Degrading Microbes Save the Bay?

Prerequisites

- **Biotechnology D11—Laboratory Techniques: Oil-Degrading Microbes** on pages 57–62

Review

- aseptic technique
- bioremediation

PARKS AND RECREATION DEPARTMENT
Eureka, California

February 28, 1998

Rosalinda Gonzales
Environmental Studies Division
BioLogical Resources, Inc.
101 Jonas Salk Dr.
Oakwood, MO 65432-1101

Dear Ms. Gonzales,

I am the Director of Parks and Recreation for Eureka, a town near the Pacific Ocean. Tourism has a significant impact on the local economy. Recently, an oil tanker spilled a large amount of oil about 10 miles from our shoreline. The oil company has done everything it can to contain and clean up the spill, but some of the oil has drifted into our bay.

Last week I was dismayed to see a great deal of oil on the surface of the shallow waters and the beach sand. In addition to being concerned for the local wildlife, I am afraid that the tourist season will begin before we are able to clean up Eureka Bay. Unfortunately, because we are at the end of a slow season, we have a very limited budget for tackling this problem. We need a quick, inexpensive, and effective way to clean up the oil without damaging the local ecosystem.

I recently spoke to a previous client of yours, who recommended your company very highly. I would like your company to help us find a way to clean up the oil without disturbing the ecosystem. I am sending samples of the contaminated water. Please keep me informed of your progress.

Sincerely,

Rebecca Childs

Rebecca Childs
Director of Parks and Recreation

BioLogical Resources, Inc. Oakwood, MO 65432-1101

M E M O R A N D U M

To: Team Leader, Ecology Dept.

From: Rosalinda Gonzales, Director of Environmental Studies

I want your department to find the most effective, environmentally sound, and inexpensive treatment for cleaning up Eureka Bay. There are a variety of ways to clean up an oil spill. For example, barriers are often used to contain a spill while skimmers are used to skim oil from the water's surface. Unfortunately, these methods are expensive and can be very time-consuming. I recently spoke with Ms. Childs, and we decided that it would be best to investigate a biological approach to this problem. One such method utilizes bacteria or fungi to break down the oil. Sometimes a cleanup crew will add microorganisms or fertilizer to enhance the growth of microorganisms already present in the soil and water.

Ms. Childs has sent samples of contaminated water and sand from the bay. Have your research teams investigate this approach by comparing the effects of adding the following treatments to the contaminated samples: fungi, bacteria, fungi with fertilizer, bacteria with fertilizer, and fertilizer. Of course, you will also need to perform a control test.

Include in your report to Ms. Childs an explanation of what to expect as the procedure is implemented and what happens to the oil in the process. Also, please add information about the food chain in a typical shallow marine environment so that Ms. Childs can see how your proposed cleanup plan will affect the local ecosystem. Finally, please keep in mind that Ms. Childs is under a very tight schedule and needs these results as soon as possible.

Proposal Checklist

Before you start your work, you must submit a proposal for my approval. **Your proposal must include the following:**

_____ • the **question** you seek to answer

_____ • the **procedure** you will use

_____ • a detailed **data table** for recording observations

_____ • a complete, itemized list of proposed **materials** and **costs** (including use of facilities, labor, and amounts needed)

Proposal Approval: ______________________________

(Supervisor's signature)

Report Procedures

When you finish your analysis, prepare a report in the form of a business letter to Ms. Childs. **Your report must include the following:**

______ • a paragraph describing the **procedure** you followed to compare the effectiveness of different combinations of microorganisms and fertilizer for degrading refined oil

______ • a complete **data table** including the microbe density and appearance of each sample for each day of growth

______ • your **conclusions** about the comparative effectiveness of each treatment

______ • a detailed **invoice** showing all materials, labor, and the total amount due

Safety Precautions

- Wear safety goggles, disposable gloves, and a lab apron.
- Glassware is fragile. Notify your teacher promptly of any broken glass or cuts. Do not clean up broken glass or spills unless your teacher tells you to do so.
- Wash your hands before leaving the laboratory.

Disposal Methods

- Dispose of all waste materials according to instructions from your teacher.
- Place all paper towels and other disposable materials in a trash can.
- Place broken glass and contaminated materials in the separate containers provided.
- Wash reusable materials such as glassware and lab utensils, and return them to the supply area.

FILE: City of Eureka Parks and Recreation Department

MATERIALS AND COSTS (Select only what you will need. No refunds.)

I. Facilities and Equipment Use

Item	Rate	Number	Total
facilities	$480.00/day		
personal protective equipment	$10.00/day		
compound microscope	$30.00/day		
microscope slide with coverslip	$2.00/day		
petri dish	$2.00/day		
test tube with cap	$3.00/day		
test tube rack	$5.00/day		
incubator	$15.00/day		
balance	$10.00/day		
50 mL graduated cylinder	$5.00/day		
sterile 1 mL pipet with pump	$2.00/day		

II. Labor and Consumables

Item	Rate	Number	Total
labor	$40.00/hour		
Pseudomonas culture	$20.00 each		
Penicillium culture	$20.00 each		
density (turbidity) indicator strip	$2.00 each		
fertilizer	$0.10/g		
distilled water	$0.10/mL		
tap water	no charge		
sterile stirring stick	$1.00 each		
wax pencil	$2.00 each		
water samples	provided		
Fines			
OSHA safety violation	$2,000.00/incident		
		Subtotal	
		Profit Margin	
		Total Amount Due	